The Complete Idiot's Guide R...

Audience Analysis

Answer these questions to understand your audience:

- ➤ What are the user's job title, job function, and job tasks?
- ➤ How frequently will the reader use the product?
- ➤ What problems is the user likely to encounter?
- ➤ How computer literate is the user?
- ➤ Is English the user's native language?

Determining Content

- ➤ Description-based = what the *product* does
- ➤ Task-based = what the *user* does

Writing Clearly

Clear writing employs …

- ➤ Active voice.
- ➤ Gender-neutral language.
- ➤ Imperatives.
- ➤ Second person.
- ➤ Present tense verbs.
- ➤ Parallel structures.
- ➤ Simple vocabulary.
- ➤ Short, direct sentences.

Scheduling

Create a realistic schedule by …

- ➤ Knowing all steps in the process from creation through production.
- ➤ Knowing how much time each step takes.
- ➤ Working backward from the target due date.

Creating Good Documents

A good technical document is …

- ➤ Usable (proven by testing).
- ➤ Accurate.
- ➤ Complete.
- ➤ Focused for the audience.

alpha books

The Five Steps to Producing a Technical Document

① GATHER

- Read all the information you can find.
- Use the product yourself.
- Interview Subject Matter Experts (SMEs).

② PLAN

- Analyze your audience.
- Schedule your work.
- Write an outline.

③ WRITE

- Use placeholders for info yet to come.
- Embed questions to reviewers.
- Make the appearance as finished as possible.

④ VERIFY

- Check your document by rereading thoroughly.
- Get feedback from SMEs and others by review.
- Test your document against the actual product.

⑤ REDO

- Correct content errors and omissions.
- Confirm your changes are on target.
- Rewrite parts needing clarification or improvement.

Finished Document

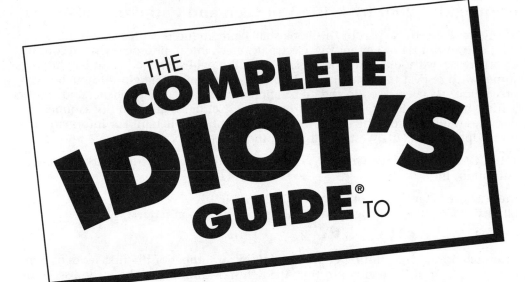

Technical Writing

by Krista Van Laan and Catherine Julian

alpha
books

A Pearson Education Company

International Standard Book Number: 0-02-864146-9
Library of Congress Catalog Card Number: Available upon request.

03   02   01   8   7   6   5   4   3   2

Interpretation of the printing code: The rightmost number of the first series of numbers is the year of the book's printing; the rightmost number of the second series of numbers is the number of the book's printing. For example, a printing code of 01-1 shows that the first printing occurred in 2001.

Printed in the United States of America

Publisher
Marie Butler-Knight

Product Manager
Phil Kitchel

Managing Editor
Jennifer Chisholm

Senior Acquisitions Editor
Renee Wilmeth

Development Editor
Jennifer Moore

Production Editor
Billy Fields

Copy Editor
Rachel M. Lopez

Illustrator
Jody P. Schaeffer

Cover Designers
Mike Freeland
Kevin Spear

Book Designers
Scott Cook and Amy Adams of DesignLab

Indexer
Angie Bess

Layout/Proofreading
Svetlana Dominguez
Mary Hunt
Lizbeth Patterson

Contents at a Glance

Contents

Foreword

A new career in technical communication can be the beginning of a fulfilling and stimulating adventure. At their best, technical communicators create information that helps people use new products effectively; they enable people to become more productive and effective in their jobs; and they help individuals use technology more effectively to support their personal activities.

For the past 25 years, I've found technical communication to be both challenging and frustrating. The field provides you with an opportunity to learn and a place to grow in your understanding of how people learn and use information. It can be frustrating when you work inside a company that doesn't appreciate the importance of communicating with customers and employees, a company that doesn't understand the value of effective information.

I have long recommended that technical communicators become experts in the user experience. By understanding the motivations of the users as they use a software or hardware product, you are better able to organize technical manuals and other technical information to ensure that they meet your users' goals.

Exactly what are users' goals, you may ask? Let's say, for example, that the users are payroll clerks. Their goals include getting out the payroll on time and ensuring that it is accurate. Their goals may focus on leaving the office by 5 P.M. Whatever the goals, effective information must be designed to help users meet those goals.

Remember that most people use our technologies simply to help them achieve their goals. They generally aren't interested in technology for its own sake. And they use technical manuals or help systems only if these tools help them get their work done, achieve their goals, and get home by 6 P.M.

Many new technical communicators get confused as soon as they get inside technology companies. They listen to product developers who want them to write about the details of their designs. They begin to believe that users really want to use the help systems and user guides they are asked to develop. They often give users too much irrelevant information and not enough critical information that will help them use products effectively to meet their goals. They describe systems and how to use them rather than real work that users want to accomplish.

The point is that it's not enough for you to learn about the product or understand what the developers are talking about. You need to focus—first and last—on the real users, finding out what they need to know and how they prefer to learn. Most of all, you need direct contact with users in their real working environments.

I've focused my 25-year career in technical communication on understanding users. From the very first project, I insisted on spending time in the users' workplace. I continue to visit customer sites, conduct usability investigations with prototype designs, and advise companies on designing and developing user-focused products and information. While the technical content is always challenging, it is always most interesting to learn about the users' experience with products and information.

Technical communicators have many opportunities to develop innovative solutions if they focus on user needs. You can mold your career on usability studies or user interface design. You can work with customer support and help create databases full of troubleshooting solutions. You can specialize in information architecture and Web delivery. You can focus on minimalist techniques to avoid flooding users with irrelevant verbiage. You can work on tools to develop effective help systems, publish on the Web, or create single-source databases. You can head into project management or even departmental management.

No matter which direction you select, you'll find lots to learn and much to explore. Welcome to technical communication. With this book, you're starting the journey on the right foot.

Dr. JoAnn T. Hackos

Dr. JoAnn T. Hackos is a leading management consultant in the field of technical communication. She advises companies on how to reduce costs and enhance the quality and usability of their technical information. She has established process standards and guidelines for conducting user studies and designing innovative information. President of Comtech Services and director of the Center for Information-Development Management, she has authored three major books in the field, *Managing Your Documentation Projects* (Wiley, 1994), *Standards for Online Communication* (Wiley, 1997), and *User and Task Analysis for Interface Design* with Janice Redish (Wiley, 1998); she is currently working on a new book on content management. She is a fellow and past president of the international Society for Technical Communication and recipient of the Goldsmith Award from the IEEE.

Introduction

It always surprises us to realize how many people who would be good tech writers think it's something out of their reach. What, *them* making a living as a *writer?* Never. Would be nice, though ....

Although you do need solid, basic writing skills, at least an average comfort level with using language, and an interest in technical topics, these attributes aren't at all rare in the general population. Now, be honest! We bet you could say they're true of you.

If so, then yes, we believe you can make your living as a writer—a technical writer. You don't need wild creativity or a burning original vision to write solid, clear technical documents. In fact, those things probably would only get in your way.

The other thing that would get in your way is simply not knowing how to start. And this book can help you with that.

What's in This Book

Technical writing is no different from many other processes: You get started, you build momentum, then you decide what to do with what you've got going. This book follows that pattern.

Part 1, "Is This the Job for Me?" orients you in the tech writing field by explaining why the market for technical writers is hot, what technical writers do, the skills a person needs to become a technical writer, and how to break into the profession.

Part 2, "Tech Writers, Start Your Engines ...," puts you on the job and gets you started updating existing documents, learning your company's product, understanding the audiences for technical documents, and gathering information. Most important, it explains the five document development steps at the heart of all technical writing.

Part 3, "Racing Toward the Finish," speeds you through completing a project from the first draft through reviews and final elements such as indexing. This also is where you'll find brief descriptions of typical technical documents.

Part 4, "Knowledge Is Power," tells you exactly what you need to know and what tools and skills you need to acquire to start out on the right foot and keep yourself on the path to professional success.

Part 5, "I Love My Job, I Love My Job, I Love My Job ...," takes a look at technical writing from several angles: new ways of working outside a traditional office, some quirks and hazards of the field that might not be apparent at first glance, and how far and what direction you might want to take your career.

At the end of the book, there's a glossary to help you acquire the vocabulary you'll use every day, along with a list of useful books and Web sites. We already know you have the initiative to find out more—you're reading this book, aren't you?

Industry Insights

In addition, you'll see plenty of insights for the unique field of high-tech tech writing:

Pros Know

These are the pearls of wisdom that make us pros and that can help you become one, too. Some are points of advice or insight; others are quotes from tech writing pros or those in other fields that tech writers work closely with. All provide inside knowledge you'll be glad to have.

Dodging Bullets

It can be easy to make mistakes, but it's easy to avoid them, too. These sidebars help you see common errors in time.

Coffee Break

All work and no play makes technical writers cranky. Coffee breaks aren't exactly play, but they are brief interruptions where we share interesting tidbits. Sure, you could live without them, but why? Everybody needs a break now and then.

Tech Talk

Every job has its own vocabulary, and technical writing is no exception. When we use a term that has a particular meaning in this field, we define it here (be sure to see the glossary, too).

Acknowledgments

Krista would like to thank the terrific documentation teams at VeriSign and Nokia NMS (especially Taro, who unwittingly uttered the words that inspired this book), Cathy for her magic touch, cats Sabrina and Dove, and everyone else who's made her work life enjoyable over the years. Most special thanks and appreciation go to her dear husband Ray Hopper.

Catherine would like to acknowledge and thank the people who, over the years, have made her work possible: the subject matter experts who endured her endless, nit-picking questions; the project team members who kept her laughing, even at three o'clock in the morning; and the reviewers who taught her to eat her documents and like it. She also would like to thank her co-author for having the idea of writing this book, and for bringing her inimitable brand of high-octane energy to bear on it.

We both would like to thank the fellow professionals who provided us with quotes, sometimes on very short notice. (You know who you are.) We also would like to thank subscribers to the TECHWR-L mailing list who allowed us to quote them in this book.

Trademarks

All terms mentioned in this book that are known to be or are suspected of being trademarks or service marks have been appropriately capitalized. Alpha Books and Pearson Education, Inc., cannot attest to the accuracy of this information. Use of a term in this book should not be regarded as affecting the validity of any trademark or service mark.

The following trademarked names have been used in this book:

Adobe®, Adobe PageMaker®, Quark®, QuarkXPress®, Adobe FrameMaker, Corel Ventura®, Macintosh®—registered trademark of Apple Computer, Inc., WordPerfect®, Corel PHOTO-PAINT™—registered to Microsoft Corporation, Palm Pilot, FrontPage®, Microsoft®, PowerPoint®, Visual Basic®, Windows®, Windows NT®, Doc-to-Help, ForeHelp, RoboHelp, Oracle, SQL, UNIX®, Visio®, Quicken, Chrysalis-ITS, IBM Correcting Selectric™, Distiller, PDFWriter, Macromedia Dreamweaver, Webworks Publisher Professional, Quadralay WebWorks®.

Part 1

Is This the Job for Me?

You can't answer that question until you know more about what the job involves; in this part, we introduce you to the field. You'll learn why the market for tech writers is hotter now than ever before, and why we think it will stay that way for years to come.

You'll also get a look at what technical writers do on a daily basis, how they fit into project teams, and what roles they might play within a company that you might not expect.

We'll tell you straight out what makes a good tech writer, too, so you can decide for yourself whether this is the job for you. (We hope it is!)

Chapter 1

The Market Is Booming

In This Chapter

➤ What is a technical writer and why should we care?

➤ The job market for technical writers

➤ Where technical writers work

➤ Why technical writing is important

Seems as if every new gadget on the market these days comes with a booklet of instructions. One tells us how to program the speed dialer on the cell phone; another how to get stock quotes and sports scores on the alphanumeric pager; another how to create a home theater around the new DVD player; and still another how to set up the kids' new video game.

And it doesn't stop there. Business today runs on software, client-server networks, and the Internet. But nobody is born knowing how to use all these tools that put the hyphen in e-conomy (at least not yet!). That's why the job market for technical writers—or "tech writers," as we call ourselves—is booming and is likely to keep booming for years to come.

If you're reading this book, you probably are either interested in becoming a technical writer, are working now as a tech writer and want to learn more about the field, or are one of the growing number whose job includes the *tasks* of a tech writer without the title. In any case, you've come to the right place. This chapter will orient you to the field and tell you about the job market and the typical tech writer work environment.

What Is a Technical Writer?

A technical writer is someone who conveys information about a technical subject, directed at a specific audience for a specific purpose. Good technical writers (you want to be one of *those*, right?) convey the information clearly, briefly, and in easy-to-understand language.

Who needs technical writers, anyway? Well, everyone. Really. We all rely on tech writers nearly every day. When we put together a piece of furniture from a do-it-yourself store, install a program into a PC, learn how to use our cell phones, or read a scientific article in a magazine, we're reading something produced by a technical writer. It naturally follows that all the industries that produce goods and market services need technical writers.

Technical writers work in many fields such as software, networking, telecommunications, semiconductors, aerospace, science, medicine, government, the armed forces, and manufacturing. Although there are jobs for tech writers in any of those fields and many others, this book focuses on the computer industry. By "computer industry" we mean a company that produces computer software or hardware.

The computer industry has become an astounding force in the world economy and continues to grow rapidly. It includes software companies such as Microsoft, computer manufacturers such as Dell, telecommunications companies such as AT&T or Nokia, networking companies such as Cisco Systems, semiconductor companies such as Intel, Internet companies such as AOL, and e-businesses such as Amazon.com. The makers of high-tech consumer products that depend on computer technologies are another part of the overall computer industry. And if you work in a company or organization that is decidedly not high-tech, but has a department or division devoted to producing software programs, then you, too, are a member of the computer industry.

Processes and methods for producing documentation are essentially the same no matter what the type of technical writing you do. Nonetheless, although some aspects of tech writing are the same across all industries, the computer industry has put its own unmistakable spin on tech writing just as it has on other business practices. We feel safe in saying that writing for the computer industry is like nothing else!

The environment is ever-changing—fluid rather than fixed and predictable. The tech writer must be flexible, or he or she will be frustrated by the lack of rules and regulations. There's no time to write involved methodologies when the marketing folks and developers are

Coffee Break

To get an idea of how fast computer technology has developed, consider this: A typical Personal Data Assistant (PDA) such as the Palm III has almost 400 times the processing speed of the Apollo Lunar Module that helped put men on the Moon in 1969. The Lunar Module's storage capacity was also small by today's standards—the equivalent of about 64K bytes.

flying by the seats of their pants. Connections of all kinds are transient and volatile as in no other business.

It's okay if you can't handle that type of work environment. You can still be a tech writer and do it somewhere that's much more organized, where life moves at a more reasonable pace. There are even software companies where the work schedule isn't frantic. Not *all* work life in the computer industry is running at breakneck speed.

But if you are up to the challenge, if you want to be part of an exciting industry and make a significant contribution, this could be for you. And it will be an exciting ride.

A Growing Community

The Society for Technical Communication (STC), the most important and largest organization for technical writers, has a membership of about 25,000, and most of those are tech writers. STC, an organization you'll hear a lot about in this book, is dedicated to advancing the arts and sciences of technical communication.

Pros Know

STC's home page at http://www.stc-va.org/ is a valuable resource for tech writers whether they're new to the field or have been in the business a long time.

Find regional and chapter information for STC at http://stc.org/.

Find the STC chapter nearest you at http://www.stc-va.org/fchapters.htm.

STC's membership includes all kinds of technical communicators: technical writers, technical illustrators, technical editors, Web designers, multimedia artists, translators, and other professionals belong to this organization. But there are many more technical writers who don't belong to STC or other professional organizations. The number of tech writers in the U.S. alone is estimated at 100,000. That's a lot of tech writers!

STC membership is divided into eight regions worldwide, each containing approximately 20 local chapters, so there's probably one where you live. We recommend you join and go to the monthly meetings. Some employers will pay professional membership dues for you, so ask. Your STC membership will enable you to network with potential employers, meet other technical writers to discuss how they do things, hear interesting presentations, and receive STC's two publications: *Intercom,* the Society's magazine, and *Technical Communication,* the Society's journal. You'll also benefit from attending the annual STC conference, which attracts technical communicators from around the world.

Where You'll Fit In

As a tech writer in the computer industry, you are likely to work in the engineering division, which gives you immediate access to the people who create the software or hardware. You also might perform exactly the same tasks as a member of the marketing department or even find yourself in an independent documentation department.

But where you fall in the corporate organizational chart doesn't always indicate what you might be writing. In some companies, the division between product documentation and marketing materials is very clear and rigidly maintained; in others, the lines are fuzzy if they exist at all. It's safe to say, though, that if you report to someone in marketing you'll probably be working on a greater variety of materials than if you are part of the engineering or programming department.

The term *technical communicator* has become popular over the past few years to account for areas of the job that are not pure writing, such as building online help, producing Web pages, and working with multimedia. We use the term "tech writer" in this book, but we acknowledge that the job is almost never confined to pure writing.

Exactly What Does a Tech Writer Do?

Of course, the specifics of the job vary widely from company to company but, broadly stated, a tech writer gathers information and then organizes it and presents it in such a way that it is understandable and useful to the defined audience. The information might be presented in book form, it might be a series of Web pages on the Internet, or it might be computer-based training provided on a CD-ROM.

Because of today's technological advances and the business trend for companies to be "lean and mean," workers are expected to wear more than one hat, and tech writers are no exception.

Often we are expected to:

➤ Know an array of software and publishing tools.

➤ Test for *usability*.

➤ Create online help and online documentation.

➤ Understand graphic layout and design.

➤ Work with printing houses and duplicating houses to handle document production and reproduction, CD duplication, binding, and packaging.

When you think "tech writer," think "jack of all trades and master of some."

Tech Talk

Usability is the practice of taking human physical and psychological requirements into account when designing programs and documents; this creates a better product and one that is more intuitive for the user. But it's not an act of altruism—improving usability reduces business costs by cutting down on the number of calls to customer support and maintains and grows market share by creating loyal customers.

Can Anyone Be a Tech Writer?

If you've ever erupted in frustration, thrown down poorly written instructions, and said, "I could write better instructions than that!" you are a good candidate for technical writing. If you enjoy explaining things or giving people directions, you are a good candidate for technical writing. Certain qualifications give you the best shot at becoming a tech writer: an ability to write and an interest in language (of course), a college education, good organizational skills, and (this is important) an interest in technology. Chapter 2, "What Does a Technical Writer Do, Anyway?" talks more about both the obvious and not-so-obvious qualities a successful tech writer needs. (The only qualifications we assume you have are an ability to write and an interest in the job. The rest can all be learned. Really!)

If you don't like to read manuals yourself—if you'd rather jump in, experiment, and figure things out on your own—you might wonder if tech writing is for you. On the contrary! Many tech writers feel the same way—and it's part of what makes them good at their jobs. You'll use that initiative, independence, and desire to figure things out every day.

Coffee Break

The job title of "technical writer" is a relatively new one. The first software technical writer wrote operating system instructions for ENIAC, the world's first large-scale, general-purpose computer, in the 1940s—but people have been doing the tasks of technical writers for hundreds of years. Books about how to identify and use herbs, called "herbals," from the 1300s, are clear examples of technical writing, as are many surviving texts from the Roman Empire.

Would you make a good technical writer?

Check the box in the "Yes" or "No" column   Yes   No

1. I love to learn about how things work.   ❏   ❏
2. When I give people driving directions, they rarely get lost   ❏   ❏
    (if they *follow* my directions!).
3. I've been told I'm good at explaining things.   ❏   ❏
4. I enjoy the English language and choosing words that work best.   ❏   ❏
5. When I read something that isn't clear, I think about how I could   ❏   ❏
    explain it better.
6. I'm able to work easily with many different types of people.   ❏   ❏
7. I'm not afraid to ask for help.   ❏   ❏
8. I have good attention to detail.   ❏   ❏
9. I can keep track of a lot of things at the same time.   ❏   ❏
10. I know tech writing is not the Great American Novel.   ❏   ❏

If you checked "yes" on seven or more of these questions, we think you can be a great tech writer! See Chapter 3, "Having the 'Write' Stuff," for details on what makes a good tech writer.

As a tech writer, you might never see your name in print (except on your paycheck!) and we can pretty much guarantee you will never see your work on *The New York Times* bestseller list. But you can have a very satisfying career, earn a good salary, help people do their jobs, and have the pride of saying with absolute truth that you make your living as a professional writer. That last statement is no small accomplishment!

Coffee Break

What's in a name? You can go by many different titles: technical communicator, documentation specialist, or information developer, to name just three. Sometimes the technical writers in a company decide they need a new job title in an effort to cover everything they do on the job. Other times, a company creates a job title to fit a certain salary grade level or department. And, there are times when tech writers feel they just want something with more pizzazz than plain old "technical writer." All things considered, we still like "tech writer" best—like good tech writing, it gets the idea across clearly and with no nonsense. Besides, other members of the product development team are just starting to get familiar with what a technical writer is and does. We're following another rule of tech writing, which is to not confuse the reader by using many different terms for the same thing.

What a Market!

We hope you're beginning to see why technical writing is a very hot job these days! Some might say it's the best of all possible worlds—it's a fun job, it pays well, and businesses are clamoring for technical writers to fill their empty job openings. You work in a comfortable office setting or even at home, and because of the nature of the job you have a high degree of autonomy. What more could you ask?

The booming high-tech industry is responsible for the current great demand for tech writers. Technology is becoming more and more available to everybody—not just a select few—and everybody needs help learning how to use it and how to get the most out of it.

Pros Know

"... [T]he Silicon Valley job market for technical writers with experience is the hottest I've ever seen, and I've been here for over 30 years. Because experienced writers are in so much demand, there's also a market for junior writers who can demonstrate anything that looks like experience. And companies are likely to be more than usually willing to take a chance on entry-level writers, provided the candidates have something that shows they know how to write and know some of the tools."

—Elna R. Tymes, President, Los Trancos Systems, http://www.lts.com

Consider how dependent we now are on software: Your grandmother might rely on software to balance her bank account and manage investments; your children are almost sure to be using the Internet and word processors to do their schoolwork; and away from work, even you might be using photo retouching programs and drawing programs to create family newsletters, invitations for parties, or flyers to help sell your house. And we haven't even touched on video games, Palm Pilots, and whatever the hot new toy will be this Christmas.

Businesses depend on software even more for mission-critical daily activities: database managers, desktop publishing programs, presentation software, cryptography to protect access to files in the midst of all this networking, inventory managers that make possible those just-in-time business practices. And there's also the hardware that enables these programs to run.

But, once again, people don't configure systems, build networks, or use programs based on psychic insight. They have to learn. And if people can't figure out how to use something, they won't. It almost could be said that a product is only as good as its documentation. So, clearly, any company that produces computer software or hardware and that wants to compete successfully definitely needs technical writers—the best they can get.

The Shortage of Technical Writers

It's no surprise, then, that the demand for technical writers has grown explosively in the United States and Canada in the past few years.

Consider just these two facts:

➤ A recent search on the Internet job site www.dice.com using only the keywords "technical writer" yielded 9,822 hits, listing jobs across the United States and internationally.

➤ Tech-writer recruiting companies are so desperate for qualified candidates that some of them have opened their own tech writing schools as a way to fill their many job openings.

At the spring 2000 STC job fair in Silicon Valley, we observed over 30 employers vying for the attention of only a few serious candidates. At one point there was only one job-hunter on the floor, and although he had less than a year's experience as a technical writer, he later received several job offers in the same week—and negotiated a salary of $20,000 *more* than what he had originally asked!

Pros Know

Posting your resumé on a job board such as www.dice.com or www.monster.com can bring a flood of results. Even a technical writer with limited experience might start receiving phone calls and e-mail within hours of posting a resumé. Be aware, however, that most of those contacts will be from recruiters combing the job boards for "hot prospects."

That's not necessarily a bad thing, though; even the small companies use recruiters to fill their openings, and they can help you get interviews. But bear in mind that recruiters are interested in getting a commission, not in finding the job that's really right for you. Keep your own goals and wishes firmly in mind—let *them* help *you*.

Of course, Silicon Valley (on the San Francisco peninsula, in California) is one of *the* hottest locations in a very hot job market, but there is similar demand in other geographic areas; we are always hearing about the newest "Silicon Valley." Your state may boast its own "Silicon Valley."

This demand enables many tech writers to negotiate perks such as high salaries, stock options, and telecommuting. Although working off-site presents special challenges for tech writers, it's no longer a hard-and-fast rule that you must live where you work. (See Chapter 23, "Office Alternatives: Working Outside the Box," for more about consulting, free agency, and telecommuting.)

Making a Living

Tech writers make a very good living, and it's just getting better. STC's year 2000 *Technical Communicator Salary Survey* shows an average salary of US$51,850 among the technical communicators they surveyed in the United States. Among Canadians, the average salary of those surveyed was CAN$49,910.

And take a look at how it's gone up in the last 15 years.

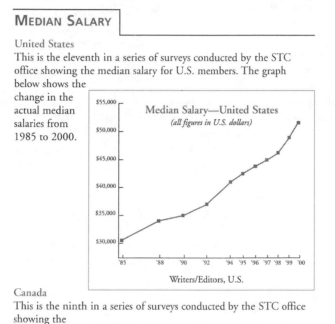

MEDIAN SALARY

United States
This is the eleventh in a series of surveys conducted by the STC office showing the median salary for U.S. members. The graph below shows the change in the actual median salaries from 1985 to 2000.

Canada
This is the ninth in a series of surveys conducted by the STC office showing the median salary for Canadian members. The graph below shows the change in the actual median salaries from 1990 to 2000.

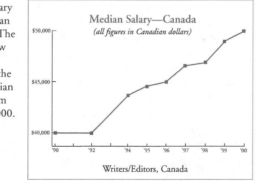

Median salary of technical writers/editors in the United States (1985–2000) and Canada (1990–2000). Used with permission from STC's 2000 Technical Communicator Salary Survey, published by the Society for Technical Communication, Arlington, Virginia.

11

Dodging Bullets

Don't expect to become a contractor, consultant, or free agent without at least a few years of solid experience in full-time, "full-contact" tech writing under your belt. The only way to earn these "wings" is in the trenches. You'll crash and burn if you try to fly solo without them.

Once you become a pro at tech writing (and that can take only a few years, with the right jobs under your belt), you can potentially earn even more as a free agent.

Although contractors working for an hourly rate can earn a lot more than full-time employees, they do sacrifice some of the benefits of a full-time job such as paid health insurance, savings plans, possible stock options, sick and vacation pay, and at least some sense of job security. But the extra income can be worth it. Peter Kent, in his book *Making Money in Technical Writing,* wrote in 1998 about freelance technical writers who made $250,000 a year!

For some contractors, free agency isn't as much about the money as about having a greater say in how they spend their days. For example, contractors have greater latitude in turning down projects that don't appeal to them (something an employee can seldom do). Even when working on-site, contractors are less often bound by a nine-to-five schedule and have more freedom to address nonwork issues during the workday. A contractor might choose to work six months of 50-hour work weeks and take another four months off to do something else. Because a contractor is paid by the hour or by the project, he or she isn't subject to the same rules a "captive" employee is.

Making a Difference

We've been talking a lot about how hot the job market is and how much money you can make. But tech writing is more than just a way to earn money. It's a satisfying job with lots of variety, which lets you use many different talents and help others at the same time. Best of all, it's a chance for continuous learning.

Technical writers create documents of all kinds, and documentation is an integral part of the product delivery. Even the best product can't overcome poor documentation—or no documentation. Customers, when polled, have said they expect to get documentation with software, and despite the trend toward providing documents only in electronic form, when customers say "documentation" they mean something on paper that they can hold in their hands.

A software CD without a manual still looks incomplete to many people—and too often they're right. Good documentation means fewer calls to tech support, and that saves money for the company (at times, *big* money—want to think about the tech support budget for Microsoft or Sun Microsystems?). So something as simple—and elegant—as a well-written user guide can have a serious impact on a division's or company's bottom line. When salary bonuses and stock options are tied to how well your company does each quarter, your job matters to many other people besides the customers or users.

Another aspect of tech writing that is less tangible, but matters a lot to some tech writers, is knowing that their efforts—their time and work—actually make a difference in someone's ability to do his or her job. For these writers, it matters that someone has one less problem in a workday or one less stress headache that week because a good writer made a critical document clear, complete, and easy to read.

Ultimately, a technical writer makes a difference in many ways—to the product, to the company, and to the user.

The Least You Need to Know

➤ Technical writers work in many different fields—in the computer industry as well as other industries—and you can, too.

➤ Tap into and cultivate your curiosity and your desire to help others understand concepts and processes.

➤ Improve your skills; companies are desperately searching for good tech writers.

➤ A technical writer's work is valuable to customers and clients as well as to the companies for whom—and colleagues with whom—they work.

Chapter 2

What Does a Technical Writer Do, Anyway?

In This Chapter

➤ A typical tech writer's tasks and duties

➤ Getting ideas across

➤ What to keep in and what to leave out

➤ Starting from scratch

➤ How tech writers fit into a company

People have a lot of different ideas about what technical writers do. The interesting thing is that many of those ideas are right! That's because technical writing covers a lot of territory, from reminders on a pocket card to multiple-volume manuals or hundreds of virtual pages on a Web site.

But there are certain things all technical writers do, no matter how large, small, simple, or complex the project. This chapter gets you grounded in these fundamentals.

Filling Some Big Shoes

By and large, technical writers are responsible for communicating information that has some specific characteristics:

➤ It's what the reader wants to know—no more, no less.

➤ The reader can find it at the moment he or she needs it.

➤ It fits in with other information the reader already knows, or it's presented with other information that helps each piece make sense and be useful.

Sound like a big responsibility? It is! Fortunately, there are tried-and-true ways to create documents that meet these needs. You'll learn about them in this book.

A Tech Writer's Day

When people think of writers, they usually imagine someone sitting happily alone in splendid and serene silence, either lost in thought contemplating a perfect turn of phrase or typing madly under the influence of the "muse of the moment."

It's true that technical writers are writers indeed and sometimes have moments like these—but there's much more to a typical technical writer's day than just writing. In fact, only about one-third of your day is spent writing if you're like most technical writers.

Your day might begin when you sit down with your first cup of coffee and check your e-mail. With more and more companies adopting a distributed model for their internal organizations, it's quite likely your colleagues have been busily working elsewhere outside your regular office hours. Keeping up with them is your first order of business.

Coffee Break

The number of technical writers in the U.S. alone is estimated at 100,000 by the Society for Technical Communication, or STC (check out their FAQ at www.stc-va.org). But we believe this number doesn't give the whole picture—people in a variety of positions perform the *tasks* of technical writers even though they're not *called* writers. Are you one?

Now is the time for you to catch up with those colleagues, check on the status of projects in the works or requests for information, answer questions, and follow up on items from yesterday. Tech writers nearly always work as part of a larger team, and that takes effort and attention.

After reading your e-mail, you probably will want to start work right away on the manual you were writing the day before. Your company's product is going to be supporting a new operating system, and you know very little about it. There's so much work to do on this document, because you have to learn about the operating system at the same time as you are rewriting parts of the manual and adding new material. This operating system is different from the one you're familiar with; although you'd like to immerse yourself in it, you don't think you can learn all you'd like to in such a short time. You enroll for a course in the new technology, and plan to delve deeper later.

Before lunch, you might spend time in a meeting to plan what documents will be needed for an upcoming new product. Because the company can't release the new item without its documentation, making sure your part is ready on time is an essential part of the "rush to market."

After lunch, you might spend some time at your desk revising a draft of a *white paper* your boss has written and asked for your help on. Editing the boss calls for tact as well as talent.

When the phone rings, it's somebody needing help—maybe the printer has a question about a manual you sent him last week, or someone from the CD-ROM vendor needs to check on a piece of art for the jewel box that hasn't yet arrived from your company. Depending on the size of your organization and how your job is defined, you might be involved in actually producing the documentation (called *production*) as well as in writing it.

Later in the afternoon, you'll spend time with an engineer or developer, learning the ins and outs of the next software application you'll be documenting. As you listen and take notes, you'll ask questions to keep the conversation headed in the right direction.

Back at your desk, you discover a stack of papers has been left for you. They are review copies of drafts you distributed earlier in the week, coming back to you right on schedule. Each one carries different comments, questions, and feedback from someone you asked to review the document. The perspectives you'll read are many; your job is to bring them all together into one polished document.

Next you get a call from one of the writers in your group. He's working on a manual that accompanies the one you were writing in the morning. Although a full-time employee, he's a telecommuter, working on site two days a week and at home the other three. This isn't so unusual for technical writers; you do it yourself occasionally.

The end of your day finds you e-mailing again—you've looked through several of the returned

Tech Talk

White paper is an industry term for a document (like a report) that states a position, or proposes and explains a draft specification or standard. It can be very technical, or hardly technical at all, depending on its intended audience. For example, a company might have a white paper on its Web site that examines obstacles associated with a particular new technology and explains why their approach to these obstacles is better than other solutions.

Tech Talk

Production is a term that stretches to include many complex processes. It might mean turning letter-perfect text files into hundreds of printed, bound, and shrink-wrapped manuals; or it might mean morphing those files into interactive pages stored in a database and integrated with scripts, graphics, and other software so they can be "published" on your company's Web site.

Coffee Break

If you're just beginning your career as a tech writer, you probably will start your job doing only as much as your boss thinks you can handle. Your first assignment might be to proofread existing documents (which is a good way to learn about the product) and to convert documents into new templates. As soon as your boss feels you're ready, you'll get real ownership of your first document.

review copies and have a few questions that can't wait. One reviewer is on the other side of the country; another is on another continent. If you e-mail them now, their answers might be waiting in the morning.

Before you go home, you think of something you want to add to the draft of the manual you are writing. You open the file and add a few notes to remind yourself of the information you want to include.

As you're walking out the door, your boss stops you and asks if you could print out some documents for a customer. Darn! You thought you could leave at a reasonable time tonight. Well, you understand that this customer is on the fence about buying an expensive system from your company, and you're happy that your documentation might be the thing to help sell the account. You print the documents and leave them on your boss's chair on your way out of the building.

Sound like a busy day? It is. And there are still many other areas of your job you didn't get to today—proofreading, creating online help, helping a non-native English speaker polish up her presentation.

Not to mention the document plan you promised your boss you'd have by the end of the week. If you work in a startup as the only tech writer, it is likely that you are also responsible for layout and design, print production, and possibly writing marketing materials. Now you see why many tech writers are trying to find a new title for themselves!

Turning "GeekSpeak" Into Plain English

Talking with that engineer or developer earlier, you were reminded of why you have a job in the first place. Engineers, developers, and other specialists, no matter what their field, usually have one thing in common: Their high level of expertise makes it difficult for them to think and talk at a level most users need to hear.

Engineers who write user documentation are so familiar with a product, they often don't realize they are leaving out crucial information. They make unconscious assumptions and assume the reader has made them, too. That can make it difficult or even impossible for a reader to follow an engineer's or developer's train of thought.

Between highly technical language and missing information, an engineer's writing often isn't the best for user documentation. Not to mention that they'd like to be left alone to do their job, thank you, which is not writing customer documentation. It's up to you, the technical writer, to bridge that gap—or yawning chasm—between how the expert says something and what your reader needs to hear. A big part of your job is acting as a translator, presenting ideas in ways that make sense to readers who don't "speak geek."

Of course, it's ideal if you can meet the expert on the same level, but that's rarely the case. Just be able to master the fundamental concepts and terms, and don't be afraid to ask questions. You can't translate what you don't understand.

Dodging Bullets

It's tempting to just nod and pretend you get it when you hear something you don't understand—but don't do it. It might make you look better at the moment, but it can have disastrous consequences later. When you encounter a term or concept you don't understand, stop the conversation and ask to have it clarified. You'll be glad you asked—and sorry if you don't.

Figuring Out What's First and Putting It There

When you're immersed in today's computer industry, you're swimming in information overload. If you feel as if you're in over your head sometimes, just think of how your readers feel. Strangely enough, that will put you right where you need to be as you begin to sort out and organize the information to go into the document (or documents) you're creating. From everything you *could* say, you must figure out what you *should* say.

It sounds simple, but a big part of your job as a technical writer is simply to begin at the beginning. What has to be done before anything else can happen? For example, before any user can install software, he or she has to put a diskette or CD-ROM into the correct drive in the computer. This becomes your starting point.

Organizing ideas seems to come easily to people who can write. You'll be surprised how much the other members of the team depend on you to supply this fundamental starting point in discussions and meetings as well as in written documents.

Writing and Maintaining Documents

These are the biggest pieces of the technical writer's pie. At startup companies, or when there's a new product being launched into the market, you'll be able to devote a lot of time to writing brand-new documents where none existed before. It's exciting and powerful to create something out of nothing.

As products continue to grow, evolve, and mature, the documentation has to do so as well. New product features and functions are added, and the documents for the product have to be updated. Often this is more a process of adding than of changing, but by no means is this always the case. Sometimes it's more important to take out obsolete information than to add in new. It's up to the technical writer to make sure a product document doesn't become a work of fiction.

Understanding How Things Work

Writers generally are expected to have mastery of the written word, but in technical writing that's not all you need to master. Understanding your company's products, their basic functions, and how they differ from each other is a key part of your job. You'll often be asked to act as the in-house "explainer" between one department and another or as a resource for newly hired employees, consultants, or contractors.

It's unlikely that anybody expects you to understand everything at the same level as an engineer, developer, or product manager. You are, however, expected to learn enough about the software application or hardware or technology so you can not only write about the product, but make informed decisions about what information is to go into the document.

Being a Catalyst for Change

An unexpected aspect of being a technical writer is that it puts you in a position (sometimes against your will) of "stirring the pot" and initiating changes in product appearance, product function, and sometimes even how a product is marketed. How does this happen?

Pros Know

"As a technical writer, it's easy to get caught up in the deliverables. For many of us, having a document, help system, Web site, or other product to show for our hard work is rewarding and significantly contributes to our professional satisfaction. But, as you work toward your professional goals, don't forget to also enjoy the processes, learn along the way, and build and foster relationships, as these aspects are key for long-term satisfaction."

—Deborah S. Ray, Site Manager, The Official TECHWR-L Site, http://www.raycomm.com/ techwhirl

Being Involved in the Design Process

Well, it's a funny thing, but when a technical writer starts asking questions about a product, people start looking at it and thinking about it in ways they didn't before. And when a technical writer brings people together from different departments who might not normally talk with each other, there can be constructive dialogue that otherwise would not have occurred. It can be very exciting to sit back at a lively meeting and realize you provided the spark that ignited all this exchange of ideas. One of a technical writer's most important functions is to spark discovery—sometimes in the most unlikely places.

Sometimes it's as simple as giving everyone their first good, clear look at what a product really does or how it really acts. Perhaps nobody realized that it takes nearly 30 steps to update the product's database until your document spelled it out for them in black and white.

The Technical Writer Is the First End User

Think about any computer industry "dream team" and you'll no doubt see visionary system architects who write their flowcharts in ink, "killer" programmers who live only to code, and salespeople who enthuse about the product even in their sleep. But who's missing? (Hint: It's someone really important ....)

The missing person is the one who ultimately will use the company's product: the end user. It's not uncommon for product designers and developers to get so caught up in *how* to make a product do something, they lose track of *why* that functionality is needed. They can lose sight of the real person who might be depending on that product to perform a task fundamental to his or her job success.

That's where the technical writer comes in. Because the user is the very person you're writing for, your job is to ask questions and make judgments from his or her point of view. The very nature of the questions you need answered makes *you* the first end user, before the product ever ships.

Coffee Break

Road signs—from "Stop" to "Main Street 2 miles" to "Lane ends, merge left"—are some of the purest, oldest, most often used, and most important forms of technical writing around. They provide just the information you need, exactly where and when you need it. Thinking about a road sign also can give you a good touchstone for judging how well a document is meeting the reader's needs.

The Least You Need to Know

➤ Remember that technical writers are part of the product development team—others depend on you, and you on them.

➤ Speak up! Asking questions is key to your success.

➤ Be ready to act as an explainer and translator as well as a writer.

➤ Embrace your role in sparking discovery and change—it's part of a technical writer's job.

Chapter 3

Having the "Write" Stuff

In This Chapter

➤ The importance of writing well

➤ Getting to the point

➤ Working with many different kinds of people

➤ Juggling as part of your job description

➤ If the shoe fits, wear it

You now have a better idea of what technical writers do, and maybe you're asking yourself whether you have the right stuff—or "write" stuff—to succeed as a technical writer in the fast-paced world of high-tech. This chapter will help you answer that question.

There's no such thing as the perfect technical writer—everyone has unique strengths and weaknesses—but some attributes can be worth their weight in gold, and the lack of them can be like carrying lead in your pockets. In this chapter, we take a look at how important (or not) writing skill actually is, and we'll bring up some other essential skills you might not have thought you'd need for technical writing.

But Can You Write?

Many people—or their bosses—think that if someone knows a lot about a particular technical topic, he or she can be a good technical writer. After all, they reason, isn't knowing the technical stuff the most important thing?

Well, not necessarily. Knowledge of the topic is only part of what makes technical writers a special breed. Although it's true you don't need to be able to write a sonnet in 30 seconds to be a good technical writer, you do need to be familiar and comfortable with the fundamentals of writing, including …

➤ The anatomy of a sentence: subject, verb, object.

➤ The appropriate *voice*—active, passive, or imperative—for each sentence.

➤ The various verb tenses: present, past, future, and so forth.

➤ How to assemble a paragraph so the sentences in it work together well.

➤ How to write sentences that read smoothly and easily.

Tech Talk

Voice is a grammatical term that describes how the subject and verb in a sentence relate to each other. Active voice means the subject is *doing* the action of the verb. Passive voice means the subject is *receiving* the action of the verb (with no indication of who or what is doing the action). Stick with active voice as much as you can so readers can see where the action in the sentence is coming from.

To succeed as a technical writer, you also need good "word awareness." This means knowing—and caring about—the difference between words such as "effect" and "affect" or "insure" and "ensure." Terms such as these, which sound similar but mean different things, can trip your readers up if they're used incorrectly. This is especially true if English isn't your reader's native language.

Word awareness also is about knowing what each word in a sentence is doing. Pick a word and ask yourself what it's there for. Does it tell you more about one of the other words? Or is it the anchor that holds all the other words in place?

But don't worry—word awareness is not something you have to be born with. It develops with practice and time. After a while, word awareness becomes automatic—like checking your speedometer while you're driving. (You do check your speedometer, don't you?)

If all this is starting to remind you of your high school English class, you're on the right track. But relax—there are lots of resources that can help you hone your writing skills and become word-aware even if you still have nightmares about "What I Did on My Summer Vacation." We've listed several in Appendix B, "For Your Bookshelf," that will get you on your way and stand by you year after year.

Pros Know

A good set of references is a technical writer's best friend. Get yourself the following and keep them handy:

➤ A good "college" dictionary, unabridged.

➤ A general style manual such as *The Chicago Manual of Style* (more about style guides in Chapter 18, "Style Guides: Not Just a Fashion Statement").

➤ A style manual designed specifically for technical documentation, such as *Microsoft Manual of Style for Technical Publications.*

Nobody can keep all the rules and definitions in his or her head or automatically know the right way to write something. But these resources can answer most of your word or usage questions.

The "Wow" Factor

If you have the aptitude for being a good tech writer, most likely you enjoy technology and finding out how things work. Many tech writers actually don't read manuals themselves when they start using software. Why? Because they like to investigate, going through each feature and figuring out what it does. When they do need the manual, however, you'd better believe they expect it to have exactly the information they want!

As a tech writer in the computer industry, you're going to have lots of opportunities to see some pretty amazing things as they're being developed. Watch an excited engineer as he or she demonstrates something that's never been done—or done as well—before. Or imagine being a writer on the team that was developing the first *GUI* or Palm Pilot—or even the first computer! If you'd like to be part of that, a career as a tech writer can give you that opportunity. There could be a lot to say "Wow!" about, and you'll be among the first to say it.

"Yes, We Have No Bananas": Communicating Clearly

In your role as translator and explainer, obviously you have to be able to communicate clearly. But what exactly does that mean? How do you know it when you see it?

Pros Know

"I got some strange looks when I got two Bachelor degrees, one in Journalism and one in Industrial Engineering, but they've really been complementary in my career as a writer. I've especially used the journalism skills. Tech writers, like journalists, need to ask those fundamental questions: What? Where? When? How? And especially: Who? And why? This is because 'who' and 'why' are what tech writing is all about—who is your audience, and why are they reading this document?"

—Krista Fritz Rogers, President, KFR Communications Associates, LLC

Clear writing leaves no doubts in the reader's mind about what's just been said. Not only are the individual ideas obvious, the relationships among the ideas also are plain. There's no question in the reader's mind about why those ideas were grouped together into the sentence or paragraph.

Writing clearly also means not beating around the bush, but addressing the subject head-on. It also means not making something more complicated than it is—or it means taking something complicated and breaking it down into parts that are easier to understand.

The most successful technical writers are those who can think clearly and organize their ideas, and then translate that clear thinking to their writing. High-tech topics are complicated enough without adding verbal complexity.

Choosing simple words over complex ones, when you have a choice, helps keep your writing clear. Organizing your information into steps helps make the topic clear for the user. Some technical writers don't even think of their main responsibility as writing—they see themselves as people who organize and present information.

We'll talk in more depth about what goes into clear communication in Chapter 19, "Writing Clearly."

"Get It?" "Got It!"

Quick—what characteristic of the computer industry makes it unlike all other industries in history? No, it's not the silicon chips. It's the pace of work and development, which seems to match the speed of microprocessors. Never before has so much activity been compressed into such small increments of time.

Changes that took other industries years or even decades to experience happen in only weeks or months in the computer industry. A new technology is introduced one day, and by next week it seems to be everywhere. Something that happened six months ago is almost ancient history in "computer years."

Because of the industry's fast pace of change, one of your keys to succeeding as a technical writer is your ability to "hit the ground running." This means being able to grasp the essentials of your company's product line quickly and thoroughly, as well as understanding what value the customer finds in the program, the piece of equipment, or the service your company offers.

The rapid pace of work also means your timeline for planning, writing, and finishing documents is going to be short. There's no time to wait for the muse of inspiration to strike when that digital clock is running.

Juggling Flaming Sticks

Perhaps the image of juggling fire is too incendiary, but the simple truth is that things will feel this way from time to time as you juggle projects, priorities, and demands on your time. A good tech writer has time management instincts and the ability to multitask. It's not uncommon for a technical writer to be working on many projects simultaneously. Those projects all are likely to be at different stages in their development, too, and each will have its own deadline—one will be in the draft stage, another might be out for review, a third might be in production, with a fourth needing an expedited update. Aside from keeping the projects themselves straight, you also have to keep track of where you are in each one, *and* meet the deadlines!

Coffee Break

Computers and the Internet are so much a part of our lives now that it's a surprise sometimes to realize how new the technology really is. It was only 1981 when IBM introduced the IBM Personal Computer and Microsoft created the DOS operating system. The Apple Macintosh and mouse arrived on the scene in 1984. In 1994, Netscape Communications was founded. Just a year later, Amazon.com started selling books online.

As a valuable company resource, you also might be writing for more than one internal department. This adds an interesting twist to your juggling, because you have to consider not only your own department's priorities but those of other departments and of the company as a whole. We did say flaming sticks, didn't we?

But juggling flaming sticks is still just juggling, and if it's done right they're no more daunting or dangerous than rubber balls. It all comes down to three simple rules:

➤ Grab the right thing at the right time.

➤ Do what you need to do while you've got it.

➤ Know when to let go and move on to the next.

The trick, of course, is knowing the right thing to grab, being clear about what to do with it, and deciding when to let go. Chapter 17, "You Want It *When?*" gives you specific strategies for making these crucial judgments every day at work.

Playing Well with Others

Whether you're part of a large documentation department or are the lone technical writer at a small start-up, you'll spend a lot of your time working with people whose jobs and work backgrounds are very different from yours. What does this mean for you? Several things. It's important to be able to accommodate other people's work styles, but it's also important to know your own work style so you can "regain your balance" after an intense work session with others on your team.

You'll also need to know how to say "no" gracefully—and how to say "yes" without throwing the barn door open and taking on more than your share—and more than you can handle.

Dodging Bullets

Pay attention to the words people use at your company and don't be afraid to ask if you don't understand the corporate terminology. The terms *book, manual,* and *document* might seem interchangeable, but in some companies each has a specific definition. It would be nice if there were a dictionary not only for business, but also for your region, your company, and maybe even the departments in your company!

We hate to say it, but it also means you'll have to pay attention to the unpleasant but unavoidable fact of office politics.

Coffee Break

Recent research has revealed that someone's "emotional intelligence" (how well they read others for a sense of their emotional state) might well have more to do with his or her career success than does a high level of cognitive intelligence (usually measured by IQ points).

Cultivating your own emotional intelligence can give you a leg up in technical writing because often it's your job to bring people together and help them work together successfully. In Appendix B, we've listed some of the many books you can buy about emotional intelligence.

Getting Along

Technical writers deal with people from all over the company; and sometimes outsiders such as consultants and contractors, trainers, and customers. It's important to develop rapport with people of all types, from the most extroverted salesperson to the shyest engineer. More than most, you will be working in what are called "cross-functional teams."

Good tech writers are able to work with everyone at the level necessary for successful communication and collaboration. Sometimes that means developing a kind of sixth sense—you have to realize that Raj doesn't want you to waste his time and Brian's abruptness isn't anything to take personally and Maria is in a bad mood, but you still have to maintain a good working relationship with each of them. The tech writer does not work in a vacuum; often you are at the center of the project, working with—and often *for*—people from all divisions of the company.

You also need to be sensitive to your own manager and his or her expectations of you. Sometimes you might think you don't have a boss because you answer to so many different people. It's easy for a tech writer to feel "bossless" when he or she is working on a promotional piece for the marketing director, a Web page for the sales team, and specifications for an engineering manager all at the same time. Despite all this, it is important for you to keep your own documentation manager informed. Your manager might expect you to work independently, but he or she is still responsible for your assignments and performance and needs to be "kept in the loop" at all times.

If you're asked to accept a project from someone other than your manager, tell that person you're happy to work on it but he or she must arrange it with your manager. Remember, if you make your manager look good (or bad), your manager will remember that when it's time to reciprocate. (Think pay raises and promotions.)

Saying "No"

There's no doubt that saying "yes" is much nicer, but sometimes you just have to say "no."

This can be trickier than it sounds. When the *time to market* is at white heat, you might find it easier to say "yes" to a request than to stop and think about whether you can really deliver. When others hear "yes" and believe it, they count on what you promised. When things fall through, it sends ripples in all directions—often with disastrous results.

Tech Talk

Time to market refers to the amount of time it takes to develop and start selling a product. Companies work to make this process shorter all the time: It can give a hearty market boost when a company's the first one out with a product—but it can backfire if the product has problems because it was rushed.

Maybe you don't have the experience to realize how much more you can (or can't) fit into an already tight schedule. If you aren't sure, ask your manager or someone with more experience in figuring schedules. You might feel really good when you say "yes" to yet another request, but if it's too much, one of two things will happen: Either you'll work all night to finish the job or you'll jeopardize the delivery.

Nobody likes to hear "no," but it's much better to tell the truth—stated in a tactful way, of course—than to let people expect something that can't be done.

The best way to say "no" is to say it straight out but follow it immediately with what you can say "yes" to. For example, when asked if you can update a 250-page draft of an operations reference guide for a customer meeting the next day, you can say, "No, I can't give it to you in one day. But I could probably give it to you in [some reasonable number of] days."

You won't always be able to respond with specifics on the spot, so be ready for those moments, too. Have a reply in mind such as, "No, I can't do that, but I'm sure we can come up with something that will work. Let's talk about it after the meeting." It's essential to convey your "no" clearly but without slamming the door in the other person's face.

Saying "Yes"

Saying "yes" is more fun and something a qualified tech writer is expected to do—often. The fact is, as a tech writer you'll have to say "yes" much more than you'd like to. You need to be flexible and willing to do a lot of things at the last minute; sometimes more things than you thought that last minute could accommodate.

Pros Know

Written communications to your colleagues don't have to be formal, stiff, or cold—after all, these are people you see every day and probably even laugh with and have lunch with. Keep your "memos" specific and focused on the topic, but also let them be relaxed and human. This keeps the doors of the relationship open and the lights on.

The documentation usually is the last thing on the developers' minds, and only when they're finished with their own piece of the project do they remember to deal with yours. This puts tremendous pressure on you at the end of the development cycle, when nearly everyone else's job is already finished.

When you do say "yes," be sure to be clear about what you're saying "yes" *to*. People love to hear "yes" and might take it to mean you'll fulfill their hearts' desires. If the marketing department wants the moon, make sure they understand that you're saying "yes" to an orbiting satellite, *not* a planetary body. You get the idea.

Also be sure to follow up your "yes" with a written memo that spells out exactly what you've said it to—including, if possible, specifics such as the date the project is due and an approximate page count if you have that. This can go a long way toward avoiding later conversations that start with, "But I thought ...."

Being Ready, Willing, and Able

Perhaps the most important qualities a tech writer can bring to the table are the qualities that assure a manager that the deadlines will be met. With schedules so important and time to market the factor that can make or break the company's bottom line, reliability is the winning quality in today's working world.

Colleagues appreciate it when you return e-mail messages and phone calls. Your boss appreciates it when he or she asks you to do something and it gets done when you say it will be. The project managers appreciate it when you anticipate problems and can tell them ahead of time if a delivery will be late. We've heard it said that writers are a difficult bunch to schedule because they seem to be more interested in turning a beautiful phrase than paying attention to the calendar. If this is you, it's something worth working on. Dependability is noticed and greatly appreciated in today's working world, simply because a remarkable number of people *don't* exhibit it.

Everybody appreciates it when you meet the deadlines. Even more than writing skills or any number of airborne flaming sticks, dependability is the best asset you can have.

The Least You Need to Know

➤ Review your writing skills—they need not be fancy, but must be solid.

➤ Learn how to multitask—it's a talent you'll need.

➤ Pay attention to relationships as well as to your projects—they're vital to your success.

➤ Cultivate dependability—it's the best asset you can bring to your job as a technical writer.

Chapter 4

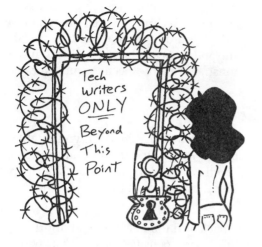

Breaking Into the Field

In This Chapter

➤ The kind of background you need

➤ The value of degrees and certificates

➤ How to kick-start your resumé

➤ Getting that crucial first interview

You're sure tech writing is for you and that you would be good at it if someone would just give you a chance. But you have no clue how to go about getting your first tech writing job because no one seems interested so far. Perhaps you have little or no writing experience and your college degree is in something like Ancient Civilizations, or you've been working in another field for years. What's an aspiring tech writer to do?

In this chapter, we give guidance on how to best prepare yourself for a career in tech writing, and then how to strut your stuff to a potential employer. It takes some work, but some of it is the same kind of work you'll be doing as a tech writer: gathering information, interviewing, following up on leads. You should be good at it.

All Roads Can Lead to Tech Writing

Luckily for you, successful tech writers come from many different walks of life. There's no agreement even among ourselves about what's the best background. Some attended school specifically for technical communications, but most did not; some have technical backgrounds, even computer science degrees, but most do not.

The good news out of this mixed bag is that if you have nothing more than the desire, the motivation, and the native ability, you *can* become a technical writer. There's no secret society ritual, no magic medal conferred by the Wizard of Oz that suddenly makes you a technical writer. What makes you one is the ability to do the job well—no more and no less. And that's something you can demonstrate to a prospective employer.

Writer or Techie? Both Can Succeed

Whether you are a college student, a new graduate, or a career-changer, you might feel as if you're facing an insurmountable obstacle: With your current qualifications, no hiring manager will read your resumé for a tech writing job. You know, it's the old Catch-22: You can't get the job without experience, but you can't get experience without a job. What to do?

Coffee Break

Would-be novelists sometimes turn their skills to tech writing to support themselves until they make it big with a book of their own. Bestselling writers who once were technical writers include Kurt Vonnegut, Thomas Pynchon, Amy Tan, and Robert J. Pirsig (author of the classic *Zen and the Art of Motorcycle Maintenance*).

Let's take a step back and look at things in another way. Successful tech writers usually started out with only one of two things: either an aptitude for writing or an aptitude for technology.

If your aptitude is for writing, be prepared to learn the technical skills that will round out your abilities. Many of today's tech writers started in another writing-related field, usually one they discovered doesn't pay well. We've known technical writers who used to be poets, newspaper reporters, and scriptwriters—all notoriously low-paying jobs that require a healthy dose of luck in addition to talent and hard work. These people, with their talent for and interest in writing and their ability to work hard in pursuing a goal, learned the technology side of the tech writing equation and became successful technical writers.

Luckily for those of you with writing aptitude and little technical knowledge, many hiring managers believe it is easier to teach technology to a good writer than to teach a technical person how to write.

If your aptitude is technology, you have an important skill to offer employers. Some tech writers we know came from technical backgrounds and never thought of themselves as tech writers at all—and still don't. Although some get into writing by choice when they discover an unexpected interest or talent for it, many others have found technical writing "thrust upon them." Maybe they were the only warm body available for the task, or the only native English-speaker on the team. Or perhaps they wrote the code and therefore were the only ones around who understood how things worked.

The Accidental Tech Writer

In many small companies, the same person who develops the software writes the manuals by virtue of being the one who knows the most about it. In large companies, the person who develops the software seldom is asked to write user documentation but often writes program design specifications. In either case, it happens that these "nonwriters" discover they enjoy writing more than programming—and a tech writer is born. People arriving at technical writing from this angle have balanced out their extensive technical knowledge by learning to write well and have became successful in a new field.

Whether you are coming to technical writing from the technical side or the writing side, as a new graduate or a career-changer, there are specific things you can do to give yourself a good foundation in your chosen career.

Dodging Bullets

Don't make the mistake of thinking that more schooling and more degrees will give you an advantage in getting a job. Because the computer industry is so young and there's no clear educational path to tech writing, work experience is more valuable than any number of degrees. If given the choice between going back to school or going to work, take the job—any job—that will give you some technical writing experience.

Building a Solid Foundation

If your strength is in writing, you'll need to shore up your technical side. As "wordsmiths" we hate to admit it, but the most important aspect of tech writing is the content. Those who understand what they're writing about make the best technical writers. Those who don't end up writing fiction even when they don't mean to.

If your strength is in your technical knowledge, beef up your writing skills to match. Content might be king (or queen), but if your sentences are garbled or your documents are disorganized, that content can't come through. It's frustrating for readers when they sense that a writer knows the topic but can't articulate key points or processes. If you don't feel confident about your writing or organizational skills, take some classes in writing or information management.

The best way to learn to write is by practicing. Try writing articles on technical topics for magazines and newsletters; these also can be used as portfolio pieces until you have weightier samples.

Master the Basics

Make sure you know to use a computer. Sound laughable in today's market? Sadly, it's not. We've seen resumés from people with no computer experience applying as technical writers, thinking their writing skills alone would get them the job. The computer is the main tool of your trade—not only because you'll depend on it for

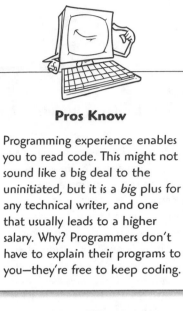

writing, but because nearly always it is a key element in what you're writing about.

Even if you don't know what sector of the computer industry you'll end up in, take computer classes in networking, programming, or both. It's certain that at least one of these subjects will play a major role in whatever product you end up documenting. The cost of such classes almost literally is like money in the bank.

Valuable programming languages to learn are C, C++, Java, and Visual Basic. Not sure what's hot in your area? Look in your local newspaper's help wanted ads and see what the job requirements are for programmers. (That not only helps you decide what programming languages to learn, but lets you know which companies are staffing up their programming department. If they need programmers today, they'll need tech writers tomorrow.)

Learn the Tools

Learn the tools. Although a lot of tech writers will tell you that your writing is important, not what word processor or graphics program you use, hiring companies feel differently—at least about inexperienced writers. Some tools are considered fundamentals everyone should know, whether those tools are used on a particular job or not. Find out what *desktop publishing* or other programs are in high demand in your area, and make sure you learn how to use them, either by taking a class or buying the software and learning at home.

Nobody cares whether you learned how to use a piece of software in a college classroom or in your brother-in-law's basement—the point is that you can do it. It is costly to buy the software yourself but there *are* ways you can learn software without paying top price.

If you're a student, look for or ask about student discounts on software. Adobe, for one, offers education discounts that can save you as much as 70 percent on the initial purchase price and still let you buy upgrades later.

Demo versions of software can be good, and often contain all a program's features but don't permit you to save a file, or disable themselves after a period of time. You can also sometimes buy used software, but use common sense if you pursue this option. For example, make sure what you're buying is a fully functioning version if that's how it's labeled, and that you receive the license and serial numbers with your purchase.

At a minimum, you should know a word processing program and the DTP used in your area. Read Chapter 20, "The Right Tool for the Job," for more about tools.

Dodging Bullets

Make sure when buying used licensed software that the seller is transferring the software *license* to you, not just the disks or the CD. A license entitles software to be used by one person at a time. Copying someone else's program files or installing from someone else's CD when they still have the program on their machine might be tempting, but it is neither legal nor ethical.

Proving Yourself

If you know you don't have the desired experience for a job you want, apply anyway and offer to take a writing test. Experienced tech writers often hate to do writing tests, but neophytes should welcome this chance to show their stuff.

What's in a Writing Test?

Many tests include simple exercises in editing and proofreading, but almost all will ask you to write some "typical" procedures. You might be asked for a procedure explaining how to tie a shoe, make coffee, or assemble a peanut butter sandwich. If you're lucky, you might get to show off with something a bit more relevant, such as writing a software procedure.

When you take a writing test, assume the reader knows nothing except how to speak English. You might even want to begin by defining key terms or by drawing a diagram and labeling the parts of things you'll talk about in the procedure. Show that you can put yourself in the reader's place.

Most important is to pace yourself so you finish within the allotted time. Your time management ability probably is an unspoken component in the test.

Showing Samples

Another strategy is to arrive at an interview with your own "writing test" already done and in your briefcase. By that we mean find a piece of technical writing you think you could improve, photocopy it, then rewrite it to make it better. Give the

Dodging Bullets

Don't use writing samples that describe how to tie a shoelace or make a peanut-butter sandwich, even though these often turn up on writing tests. A potential employer is much more likely to take you seriously if you show work that was—or actually could have been—used by a paying customer in a realistic industry setting.

Tech Talk

Open-source software (OSS) is software in which the program source code is openly shared with and among developers and users. The idea behind open-sourcing code is that if it's readily available, people (programmers) can see how it works and will choose to customize programs to work with it. They're also expected to share their work with everyone else, benefiting the entire programming community—thereby also benefiting users.

interviewer a copy of the original piece and your rewrite. (Make sure your name and contact information are on both pieces.)

This can be a good way to show what you can do and demonstrate your ability to think ahead. Be prepared to discuss why you chose that piece to rewrite and how long it took you.

Building a Portfolio

Don't expect that a hiring manager is going to know intuitively what a wonderful writer you'll be and hire you with no experience and no proof of writing ability. In fact, most interviewers will *require* you to show writing samples.

What do you do if you have no writing samples? That's easy: Make some! Here are some ways you can build your portfolio:

➤ Do volunteer work. Plenty of nonprofit organizations need help with their newsletters or other writing tasks.

➤ Write documentation for shareware and freeware programs.

➤ Write documentation for the hot new *open-source* operating systems and applications. Developed and distributed free of charge, systems such as Linux and programs that run on it create a huge need for user and programmer documentation.

You probably won't get paid for any of this work and it might never be read or used, but you'll benefit by having a solid portfolio to show to potential employers.

Recast Your Resumé

Make sure your resumé is written in a way that improves your chances of landing a job. That means emphasizing every bit of writing experience you can lay claim to. If possible, avoid listing jobs or projects that don't relate to the work you are seeking. Recast your resumé so the jobs and projects you do include sound like technical writing or show a skill that is transferable to tech writing (without lying, of course).

An obvious tactic that nevertheless many writers don't use is to write your resumé as a technical writer should: effectively, succinctly, completely. Your resumé needs to show off your abilities. And for heaven's sake, make sure you run the spell-checker! (You wouldn't believe how many applicants for tech writing jobs send out resumés full of silly errors.) Then, have someone else read it, preferably an experienced technical writer who might be able to help you present yourself in the best way possible, or who can at least point out problems you might not see.

See Yourself as a Professional

Self-imagery is powerful: Visualize yourself as a tech writer as you write your resumé. And, really, after you follow the guidance in this book, you essentially *are* one: You have writing samples that were read by real users, you have experience in the tools of the trade, and you have an understanding of the process of producing user documentation. You're there!

It's important to follow two basic rules that aspiring tech writers often violate:

➤ Never start with a plea for someone to train you. Instead of leading off with "Objective" as a heading, try using "Summary" instead, and write something like "Junior technical writer with experience in …" then list the types of writing you have done.

➤ If you have learned the basic publishing tools in demand in your area, say so up front. Use a heading such as "Skills" or "Software" after the summary. Under this you can describe your software expertise, perhaps in a bulleted list.

Making the Most of What You Have: For the Career Changer

Before deciding you have to bail out of your current job, see if you can turn it into a technical writing job. Or if you work in a company that has a technical writing department, tell the tech writing manager about your interest. Ask if there's anything you can work on as a way to get your feet wet.

Don't assume management will be opposed to the idea, either. At two different companies we've worked in, administrative assistants who wanted to be tech writers managed to transfer into the documentation department. Management is always pleased to acquire technical writers, especially those who already know the company ropes.

Kim Reiter
1234 Macmillan Dr., Anytown, CA 99999 USA
408-555-1212

Objective: An entry-level technical writing job where I can
learn new skills and build my professional experience.

Education: BA in History, University of California - Davis,
1993

Work History:

1997-Present
Executive Assistant, A
Manage appointments
do typing and filing, a
staff meetings.

1993-97 Administrati
Geological Surveys, D
and typed letters for th
technicians. Also help
clients.

1991-92 Copy editor
Evelyn," *Eye in the Sk*

1990-1991 Waitress,

1989 Cashier, Jones M

Hobbies:
Gardening, bowling

KIM REITER

1234 Macmillan Dr.
Anytown, CA 99999
(408) 555-1212

email: kreiter@anytown.org   WWW: http://www.anytown.org/~kreiter.htm

---

SUMMARY

Junior technical writer with 8 years of business and marketing experience. Published author.

- ✓ FrameMaker, Microsoft Word
- ✓ NT, Windows
- ✓ FrontPage, DreamWeaver
- ✓ Vision, CorelDRAW

WORK EXPERIENCE

Acme Gadgets & Gizmos, Oakland, CA May 1997-Present

Writer / Executive Assistant, Marketing
- Responsible for researching, organizing, and writing weekly and monthly sales reports and quarterly marketing strategy reports used in executive staff meetings.
- Coordinated with printing vendors to revise and produce printed marketing materials for regional sales reps.

Crenshaw & Mellon Geological Surveys, Davis, CA Sep 1993-May 1997

Report Preparation Specialist
- Assisted three survey engineers to prepare monthly and quarterly survey progress reports, including supporting spreadsheets, bar graphs, and charts.

WRITING EXPERIENCE
- Copy editor, *Eye in the Sky* campus newsletter, UC-Davis. 1991-92
- Magazine article "The Role of Ikat Weaving in Ancient Persia," published in *Threads* magazine, August 1994

EDUCATION
- University of California, Davis, CA. BA History, 1993

PROFESSIONAL MEMBERSHIPS
- Society for Technical Communication
- International Association of Business Communicators

— References available upon request —

This before-and-after example of a career-changer's resumé shows what you can do to improve your job-hunting chances.

Coffee Break

Controversy abounds about whether certification should be required for technical writers as it is for engineers and accountants and other professionals who have to take a test to prove their ability to perform their jobs. If tech writers were certified, the argument goes, a hiring manager could be sure of minimum qualifications. The anti-certification side contends that the field is so diverse no standard certification would be meaningful across all industries. Some even assert that any hiring manager worth his or her salt should be able to judge a candidate's qualifications whether certified or not. And so it goes.

Keep your eyes open for opportunities. One of the best tech writers we know was a bank teller interested in computers. When a writing job opened up at the bank, he talked the manager into hiring him. After two years of writing at the bank, he moved to a software company where after a year he is earning twice his bank salary.

You can use your previous job experience to your advantage on a resumé. Don't think you have anything tech writing–related? Think again.

Pros Know

The Open-Source Writers Group is a nonprofit organization whose primary goal is to improve the overall quality and quantity of free open-source and open-content documentation. Their Web site at http://www.oswg.org/oswg/ includes a page where you each register as a volunteer writer, editor, or proofreader for documentation related to open-source projects. This is a great way to acquire samples for your portfolio and sharpen your skills.

Anything in either the writing or technical arena is going to help you in your new career as a tech writer. Written personnel guidelines? That's procedure writing. Helped your boss edit presentations? Depending on the subject matter, that might have been technical editing. Developed clean, accurate, and well-labeled flowcharts? That's related to both document design and procedure analysis. Embedded clear and thorough comments at strategic places in a program? That's only one step away from tech writing. Taught English? That's a double hit: Teaching is a terrific background for the field, and technical writing has a lot of English majors. Used a word-processor or graphics program on the job? Emphasize that.

Write down everything you've ever done, on any job, that relates to technical writing, from flowcharting to writing reports. As you continue reading this book and learning more about the field, you'll find more to add to your list. You might be surprised to realize that some of your previous work was indeed technical writing.

Let's Discuss Degrees

Yes, having a college degree (any degree) definitely will make it easier to get a job as a tech writer. That said, you might think that degree should be in technical communications. If your degree isn't in technical communications, maybe you're considering returning to school to get one. An "education" in technical communications can be anything from a six-week certification program at a night school to a Master's degree from an accredited university. Do you really need it?

Our short answer is no, you don't need to have a degree specifically in technical communications. Our long answer is that a college degree is important, and one in technical communications does have its advantages, but is by no means essential for becoming a technical writer. If you're only starting to plan for college, by all means, major in technical communications if that's what you know you want as a career. But if you already have a degree in any field, there's probably no need to go back for another one. We know technical writers with degrees in such wide-ranging subjects as art, medieval literature, and nursing, along with the more likely journalism and engineering. The point of the degree is that it shows that the writer knows how to learn, and how to stick with something—two things every tech writer must do every day.

Once you have a degree, if you still want to go to school for technical writing, we recommend one of the short certification programs. These programs give you a quick overview of technical writing, teach some tools, and often help you assemble a portfolio to use for job hunting.

Unlike many college and university programs, schools offering certificates (most of which have nothing to do with "certifying" a tech writer; they are more like vocational schools) use professional, employed tech writers as instructors. Think you'd rather be taught by Ph.D.s? Think again. These working instructors give you insights about on-the-job reality, not textbook theory, and can be valuable contacts later. Your fellow students, too, will be good contacts for the future. As each of you finds a job in the field, you will be able to help each other (yes, this is networking).

When investigating a school get specific, concrete answers to these questions:

➤ Do this school's graduates find work? Make sure you speak to people who have attended the school. Find out if they believe being a graduate helped them get a tech writing job and if they use what they learned at the school.

➤ Are the instructors working technical communicators? You don't want a teacher who never heard of the subject before he or she was told to teach it, or one who has been out of the business world for many years.

Pros Know

The STC maintains a list of technical writing schools at http://www.stc-va.org/facademic.htm. Explore your options before deciding on a particular school or program.

➤ Will you write at least two different types of documents that can be used as portfolio pieces? One school we know of has the students write and format a hardware manual and a software manual. The manuals are based on real products, and the students end up with professional-looking work samples.

➤ Will the school help you find a job? Check with their placement office to see what their record is in placing their graduates and confirm that there are more than one or two companies hiring their alumni.

If a school's answer to any of these questions is unsatisfactory, scratch it off your list unless it's the only one in your area. Even then, think carefully before you invest your time, effort, and money in their program.

Why Is It So Hard to Get an Interview?

With all the open jobs and desperate employers, you'd think it would be a cinch to get interviews and step into that first tech writing job. Yet, even though you're motivated, you can write, and you have a great-looking resumé, maybe you're still having trouble getting to square one. If the job market is so hot, and managers are so desperate, why aren't they beating a path to your door?

Take heart. It's not necessarily you—managers don't have some sixth sense enabling them to look at a resumé and know who can or can't do the job (which is too bad, because then they would look at *your* resumé and know you *could* do it).

The cold fact is that hiring managers are under enormous pressure to produce, and the image in their minds of the employee they want looks more like a hired gun than a kid with a slingshot. And even though ultimately it's how good a sharpshooter you are, and not the weapon you start with, many managers need help seeing past their ideal.

43

And it's also a cold fact that people with no experience are a dime a dozen. Most managers are holding out for somebody with at least a little extra something to offer beyond eager ignorance. Nowhere is it more true than in the computer industry that time is money, and managers do not look with favor on that inevitable learning curve. In fact, they often wait months looking for the "perfect" candidate, when they could have spent those months training an eager and competent newcomer.

How can you turn this to your advantage? Become that someone with the little something extra—extra software or hardware knowledge, extra samples in your portfolio, extra writing projects you've created or taken on at previous jobs, extra time spent writing as a volunteer on open-source applications, extra effort in following up on that ad or job posting, or in pursuit of that elusive interview.

And—here's your ace in the hole—be willing to work for a reasonable rate. We don't mean for peanuts, but for a salary that reflects your level of experience. Merit increases can be regular and generous in the computer industry, and bonuses and stock options are available in many companies.

Even if the offer you get isn't the number in your dreams, don't forget that you're still getting something that no amount of money can buy: experience. In six months or a year, you'll be a tech writer with that much experience under your belt—and that's a whole new ballgame.

Use a Recruiter

No matter what you might hear about recruiters (they're only in it for the money, they don't care about the job hunter), the fact is that recruiters have access to a lot of hiring managers and a lot of jobs. Recruiters, because they make their living placing people, often will work to help a talented newcomer find a job.

Recruiters advertise in the "help wanted" section of the newspaper, on the Internet, and in the Yellow Pages. One sure way of finding a recruiter is to post your resumé on a job site such as http://www.dice.com. Prepare your resumé as we suggest in this chapter so the recruiters' scanners will pick up the keywords they're looking for, such as *technical writer, FrameMaker,* and *RoboHelp.*

Network, Network, Network

It's your responsibility to make sure you meet as many working tech writers and hiring managers as you can. This puts the law of averages to work in your favor. The more people you talk with, the better your chance of meeting the manager who will realize that you have the desired qualities of a tech writer (discussed in Chapter 3, "Having the 'Write' Stuff").

How do you meet real, live, working tech writers? Network, network, network!

➤ Go where the tech writers are—join the STC and other professional groups and *go to meetings!* Get involved in a special interest group and serve on committees.

➤ Subscribe to tech writer mailing lists and newsgroups online, as well as print magazines.

➤ Let everyone you meet know you're looking for a job as a technical writer.

Go to the STC Conference

The employment corner at the annual STC conference is an excellent place to meet potential employers and submit your resumé. If you go to the conference, be sure to stop by there—but that's not where you're most likely to strike employment gold. There are many, many newcomers, all eager to break into the field, and hundreds of them are at this conference trying to find work. It's easy for one more to fade into the crowd.

It's easy, that is, unless you engage those good tech writer brains to stand out from the crowd and work the conference opportunity to your best advantage. First, make sure you look professional; this is a trade conference, not a vacation. Dress as you would to make a good impression on your first day of work—then carry yourself like you mean it and believe it.

Do what you can to meet the people who are in a position to hire. Do something (pleasant, please!) to help them remember you. If you spend the whole STC conference attending "Lone Writer" or "How to Write Your Resumé" sessions, don't be surprised when you leave the conference still alone, with a great-looking resumé that nobody has seen. Do what successful beginners have done to make yourself stand out at the STC conference:

Pros Know

"Some people join the STC just to receive its *Journal* and the *Intercom* magazine, and perhaps a local newsletter. Others join because they get a discount on local events, such as seminars, and monthly meetings, or regional and international conferences. Many join to network—to meet and interact with other technical communicators, to get the scoop on local recruiters, employers, managers, college courses, and job openings, as well as to keep up with the changing technology."

—Guy K. Haas, President, Silicon Valley Chapter of STC, Staff Technical Writer for Selectica

Coffee Break

Don't simply send out resumés. Meet the people you're interested in working for and with. It's much harder for someone to say no to a person he or she has met face to face, and it gives you both a chance to see if there could be a working rapport. And don't forget the value of the informational interview, which has a long history of getting people's feet in the proverbial door.

Pros Know

A mentor can be an invaluable aid to a beginning tech writer. The mentoring program on TECHWR-L, the official Web site for the TECHWR-L (technical writers') mailing list, helps match students or beginning writers with mentors. Go to www.raycomm. com/techwhirl/mentoring/ and find out how you can request information on mentoring. Read the rest of the TECHWR-L Web site at www.raycomm.com/techwhirl/.

➤ **Attend the sessions aimed at managers.** Go where your next boss might be. Sessions aimed at managers are clearly marked on the program.

➤ **Don't be shy.** Many technical writers seem to be introverted, or at least reserved, and often it's difficult for them to strike up conversations with strangers. If this describes you, however, remember that being able to talk to a lot of different people is a big asset for a tech writer on the job, and this is your chance to practice that. So, uncomfortable as it might be, do it. For starters, force yourself to speak with, say, three to five more people than you really want to at each session. It does get easier with practice.

➤ **Be who you want to be.** Think of yourself as already a technical writer, not a "wannabe." Dress and act like a professional and carry business cards with your name and contact information and the title "Technical Writer."

➤ **Follow through on the details.** Remember that a technical writer is responsible for following up on myriad little details, each of which is important to the whole. Pay attention to the details of how you present yourself: your cards, your appearance, your resumé. Don't make the mistake of having a beautiful resumé and good samples but being 20 minutes late for an interview or a get-together dinner at the conference. All the parts are equally important when you're trying to get that break.

➤ **Make a lasting impression.** Make sure potential employers remember you. Collect cards from the managers you meet and follow up with a friendly note via e-mail right after the conference, asking them to keep you in mind if an entry-level opening comes up.

Winning Interview Tips

When you do get an interview, remember to act like a tech writer: Go to the interview prepared by knowing as much as you can about the company and the job. Be aggressive in your efforts without being annoying. Ask intelligent questions. Show an interest in how things work (at the company where you want to work, that is).

Especially remember to look professional while job hunting. We know we're repeating ourselves, but there's a reason for that. You might have the impression that everybody in the computer industry wears cut-offs, a company T-shirt, and sandals. Not so. Or at least not on the first day—and probably not the day they were first interviewed. Although it's true that "business casual" is more appropriate today than dressing like the Secret Service, it never hurts to look just a little bit better than you need to.

Andreas Ramos, documentation manager and chairman of the National Writers Union's technical writer's trade group in Silicon Valley, posts the following advice on his Web site at http://www.andreas.com:

➤ Managers like you to ask questions; it shows you're bright, interested, and paying attention. If you don't ask questions, they'll think you're either too dumb to do the work or not interested in the job.

➤ On a job interview you're nervous. When you're nervous, it's hard to remember what you want to ask. So write down your questions beforehand. Always bring a nice-looking notebook and take notes on what the manager tells you (they like that). Because your notebook is already open, it's easy to refer to the written questions you want to ask.

➤ Research the company before the interview. Go to their Web site and print the relevant pages. Read them again while waiting in the lobby for the interview. If you know anyone who has ever worked there, talk to them.

A special reminder to you reserved soon-to-be tech writers: Remember to try to close the deal. Make it clear at the interview that you want the job, and follow up with a written or e-mail note or a phone call. All you need is for one manager to say yes.

Persistence Pays Off

It sounds obvious, but persistence does pay off. Everybody we know who has really wanted to get into technical writing has done so. It seems clear that the people who get the jobs are the ones who want them a lot and act—persistently—on that desire.

If five people apply for a job and their qualifications are roughly equal, the one who sends a thank-you note and takes the initiative to call a

Dodging Bullets

In following up on an interview, we're not suggesting you make a pest of yourself with too-frequent phone calls or e-mails. Once a week is enough unless the manager explicitly asks you to call sooner or more often. Calling every day is a great way to make sure you *don't* get the job!

few days or a week after the interview to touch base, express continuing interest, and ask for a status check is most likely to be offered the position. It's hard to turn down someone who expresses a sincere interest and shows strong motivation.

Pros Know

We aren't going to tell you how to act at an interview; there are plenty of books about that. But we must mention one important thing that many interviewees neglect to do: Tell the interviewer how much you want the job. It's harder for an interviewer to say no to someone who really wants a job and asks for it. This also lets the manager know you didn't come to the interview merely on a "fishing expedition."

Bear in mind that the way you show your persistence to the hiring manager demonstrates how you'll show it on the job. Persistence is an important characteristic that makes a good technical writer—but so is tact. Managers want tech writers who will doggedly go after information until they get it—but in a nice way.

Sometimes it takes a while to find the person who will give you that first tech writing job, even in a hot market. But the demand definitely is out there. Don't give up, and do "work smart." Keep going to interviews with the right background, the light of motivation in your eyes, and the look and attitude of a professional. Eventually, an interviewer will realize that you've got what it takes to do the job, and do it right.

The Least You Need to Know

➤ Don't obsess about whether you have enough or the right education; a degree does help, but nearly any degree is acceptable.

➤ Do "volunteer" tech writing to build your portfolio and rack up experience.

➤ Polish your overall presentation—appearance, attitude, resumé, even business cards—to improve your chances.

➤ Keep at it! The big break will come.

Chapter 5

What Makes a Good Document?

In This Chapter

➤ The elements of a good document

➤ How to judge documents

➤ Understanding document categories

In this and other books about technical writing you'll see references made to "good documents." But how do you know if a document as a whole is good, mediocre, or a failure? You can't do that if you don't understand what makes good documents—of any type—and what distinguishes one type of document from another.

Documentation is any reference material that supports something else—in our case, a software program or piece of hardware. This collection of information might be in an electronic file or printed on paper. It might be a programmer's notes embedded in a program or a full set of beautifully bound books. It may be disorganized and incomplete, or 70 volumes of clear and essential information—it's all documentation if it supports a product.

A good document is ...

➤ Accurate

➤ Complete

➤ Consistent

➤ Clear

➤ Useful

➤ Attractive

This chapter acquaints you with the basics for making informed judgments about these characteristics.

"A" Is for Accuracy

What counts more than accessibility, crisp writing, and a knockout layout? Correct information, of course! Your reader trusts that the document's information or instructions are indeed fact and not fiction. Nothing shatters that trust faster or into smaller pieces than your reader discovering—always too late—that a crucial piece of information, confidently accepted, is wrong.

But user frustration can be the least of your worries. Timelines, budgets, and even entire projects can be put at risk. More than one major consulting contract has been lost over one too many factual mistakes in the documentation.

As a technical writer, you might feel as if you're in a tight place when it comes to accuracy. After all, you didn't write the code or design the product, right? But the documentation is what users can see and point to, and point they will. What to do?

Here's how to be sure your documents receive an "A" for accuracy:

➤ **Know what you're writing about.** The best first line against inaccurate information is you, the writer. Refer to Chapter 8, "Learning Your Topic," for information about learning the product.

➤ **Keep thorough and organized notes** when you gather information and double-check (even triple-check) essential facts, figures, parameters, and other core specifics. See Chapter 10, "Gathering Information," for more about gathering and organizing information.

➤ Get your drafts reviewed by people whose review is meaningful—people who know, on their own, whether what you've written is correct. Chapter 12, "Everybody's an Editor," guides you in choosing and working with reviewers.

Pros Know

Relying on others to check the accuracy of what you write is only common sense (you can't know everything), but there's no substitute for direct experience. Jump at any and every chance you get to use or handle your company's product or to meet your company's customers or users. The insights you'll gain from an hour or two of real-life interaction will be invaluable.

➤ Require formal reviewer sign-offs with signatures and dates during development (not just at the end). This reminds the rest of the team that they also are stakeholders in creating accurate documentation.

➤ Test your documents for correctness before they're released—there's nothing like a tough road test to bring out the squeaks and rattles. There's more about document testing in Chapter 15, "Making the Final Laps."

Completeness Counts

Accuracy and completeness are different things, but they go together like "coffee" and "cup." If there are 10 steps in a process and you clearly, correctly, and brilliantly describe only eight of them, your readers won't thank you. Providing complete information is every bit as important as being accurate.

Make Processes, Documents, and Sets Complete

Completeness counts not only in processes but also in the document as a whole, and in sets of documents. That's why planning—from insert sheets to multi-volume sets—is so important.

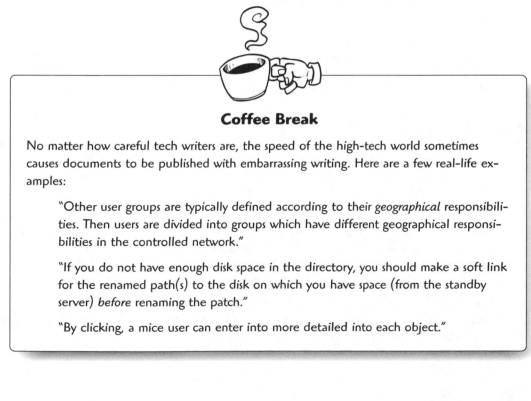

Coffee Break

No matter how careful tech writers are, the speed of the high-tech world sometimes causes documents to be published with embarrassing writing. Here are a few real-life examples:

"Other user groups are typically defined according to their *geographical* responsibilities. Then users are divided into groups which have different geographical responsibilities in the controlled network."

"If you do not have enough disk space in the directory, you should make a soft link for the renamed path(s) to the disk on which you have space (from the standby server) *before* renaming the patch."

"By clicking, a mice user can enter into more detailed into each object."

To be complete, a document—or document set—must contain the information the user needs. It's not enough to write a procedure on how to install in an NT operating system if half your users have UNIX operating systems. It's not enough to give users an overview of a procedure if they need minute detail. And it's certainly not enough to give users a recitation of what the features and menu commands do—and then fail to tell them how to use them in their workflow.

Completeness Includes Editing and Proofreading

There are other ways a manual should be complete as well. Finishing touches such as editing, proofreading, correct headers and footers, and indexes are among the quality items that make a document complete.

Check for Completeness Early and Often

The best time to check for completeness is at the same time you check for accuracy. Include a statement about completeness in your formal sign-off sheets, too. That way you'll know you're always on track. There should be no surprises when you're ready to test your documents.

Clarity Is in Good Writing

If your writing isn't good, the other qualities don't matter much. An effective document must be well-written. No matter how much you know about a subject, it isn't enough if you can't get your ideas across. Technical writing is not just about knowing, it's about explaining. To do that, you need to be able to write clearly. We talk about some of the rules of good writing in Chapter 19, "Writing Clearly."

Consistency

In conventional English, there is an abundance, a plethora, in fact a myriad of synonyms—one and a half million in common use. One version of *Webster's Dictionary* gives 13 million uses of its three-million-word collection!

What this boils down to is that the same, or nearly same, meaning can be expressed with a wide variety of different words. Isn't this a good thing? Not really. Anytime you use a different word or term to refer to something you've already called by another name, you risk confusing the reader. Is this new term a different object or action? Does it mean that the previous object has somehow changed into something else?

Even if the reader doesn't become confused, he or she has at least had to stop and assess whether this new term means the same thing as the previous one. You can only hope the conclusion is the correct one.

A good document is consistent in its use of language, in its use of iconography, and in its use of typographical conventions. The user should never have to wonder about the significance of a word's use. Typographical conventions to indicate actions such as user input or system messages should be consistent. Even if you don't explain the conventions in an introductory section, the conventions should be so consistent that the user's subconscious mind understands their meaning. (We discuss consistency in more detail in Chapter 17, "You Want It *When?*")

Having a Handle on Usability

"Usability" is a term you've probably heard a lot, but you might not know exactly what it means. It's the capability of being used easily and effectively by people.

Knowing Usability When You See It

Usability of a product boils down to five questions:

➤ Is it easy to learn?

➤ Is it efficient to use?

➤ Can a user recover quickly from errors?

➤ Is it easy to remember what to do?

➤ Is it fun to use?

The usability of your document is the measure of the quality of the user's experience when he or she is interacting with your documentation. To determine a document's or Web site's usability, you really have to ask only one thing: Does my document make it easy to find the information the user needs?

Documentation plays a very important part in the usability of software. If the software has less-than-optimal usability, the documentation must try to make up for it—a big responsibility for the tech writer!

Dodging Bullets

Over time, you're bound to find yourself favoring one reference book over another or working from one style guide more than another. But don't fall into the trap of becoming rigidly attached to a single way of doing things. There are many roads to reach the same place, and insisting to your team that only one way is the "right" way is an expressway to Friction City. Be open to taking the scenic route occasionally, even if it's not your first choice.

Pros Know

Your copy of *Joy of Cooking* is a perfect example of good technical writing practices. Its recipes are short, clear, and accurate, and use the active voice to deliver instructions succinctly. They also contain exactly the information you need when you need it.

Coffee Break

When you start looking into us-ability, one of the first names you'll encounter is Jakob Nielsen. A Distinguished Developer at Sun Microsystems until 1998, his work in relation to the Internet and World Wide Web is at the core of the growing field of usability. His is a name you need to know.

Pros Know

Check out these Web sites for lots of good information about usability:

http://www.usabilityfirst.com/
http://www.usableweb.com/
http://world.std.com/~uieweb/
biblio.htm

Time and Money Well-Spent

Many companies devote a great deal of money and effort to usability testing for their software. This can involve anything from hiring a consultant to do a *heuristic* study (a method of solving problems by intelligent trial and error), to videotaping users as they attempt to perform tasks. Lucky tech writers get their documents usability-tested, too; this usually means that someone goes through all the procedures, ascertains that they work, and evaluates areas that are less than clear.

A Natural Career Path for Tech Writers

Many tech writers gravitate toward the field of usability. It's a natural step for many writers who become involved with improving the product's usability; the writers are naturals for writing test procedures.

At the core of document usability is information transfer, which means moving information from where it is to where it needs to be (for example, from a book to a brain). The purest function of a technical writer is to facilitate information transfer. But keep in mind that the user isn't interested in a document as a whole or in all the individual pieces of information that are contained in a document (in fact, those pieces are just in the user's way). The information the user wants to transfer is that one particular piece that he or she doesn't have, but needs—a friend's phone number in a distant city or where on the Web he or she can order fruitcakes in July.

Keys to Improving Document Usability

The following points will help you build in high usability as you write.

Even if your reader is learning or doing a complete process, he or she still needs to go one step at a time. Your document must make those steps clear, and they must be easy to identify as the reader switches back and forth between reading and doing.

The reader also needs to be able to determine, at any time, which procedure that step is part of (if there are multiple, similar procedures). These issues relate not only to information organization but also to visual layout elements that contribute to usability.

In addition to organization and layout, give some special thought to *entry points* where your reader first encounters the document. Where's the first place a user looks? In print documents, it's usually the index, with the table of contents not far behind. With Web-based documents, it's usually a site map or linked index, or a site-specific search engine or word search within the document. These are your first chances to help the reader—or give him or her a headache.

If the reader is browsing for general information, she probably looks at the table of contents or site map. If the hunt is on for a specific topic, she probably looks at the index or search engine. But no matter what entry points your user starts with, the most important factor in usability is that she is able to easily find the information she is looking for.

Don't know your audience? Then go get a cup of coffee and read Chapter 9, "It's All About Audience." You can't design a usable document until you have a clear idea of what a specific group of people (your audience) needs to see and wants to know, and why.

Pay attention to your own internal sense of what might be lacking from a document or collection of documents. Often technical writers have a better view of the big picture than others and can better spot where pieces are missing. Programmers and designers sometimes are a little too close to the project, and members of the sales force often are one too many steps removed.

Tech Talk

An **entry point,** in programming jargon, is a place in a program where execution can begin. Expanded to apply to documentation, an entry point is any point or avenue a reader can use to access a body of information. An index is a collection of entry points, as is a table of contents. Subdivisions of a document also are entry points.

Dodging Bullets

Resist the temptation to reinvent something fundamental, such as presenting the table of contents as a decision tree, for example. Readers diving into a whole book of new information have all they can handle. A clear and orderly table of contents might look boring to you, but to them its familiar structure is an island of the known in a sea of the unknown.

Will It Fly? Judging Documents

Usability, accuracy, and completeness cover the big points, but there's still more that goes into making a document good. The Society for Technical Communication holds an annual international competition for technical publications, online communication, technical art, and video. Entries in the technical publications competition are judged on production, design and typography, copy editing, content and organization, and graphics. A winning publication must also meet the stated purpose for the intended audience.

Pros Know

"Technical writers need to be usability-literate because our work is often the first support users consult when they have problems with a product. Data about how people use products and documentation can guide our decision-making about document organization, style, and content. ... All who contribute to developing products can benefit from the usability knowledge our profession has helped to build."

—Stephanie Rosenbaum, President, Tec-Ed, Inc., http://www.teced.com

To get you started, here are some fundamental criteria to use in judging a document, whether it's one you're writing or have written, or one you are simply evaluating as a resource. Most of them are simply common sense.

➤ Usability, accuracy, and completeness are the Big Three; if a document lacks any of these, it can't be good.

➤ The document should be attractive to the eye—or at the very least not be actively offensive. Readers resist reading what they don't like to look at.

➤ Skim some paragraphs throughout the document with attention to paragraph length, sentence length, and word choice. Information should be presented in bite-sized chunks with one clear idea per chunk. Simpler words always win out over more complex ones.

➤ Hold a page at arm's length, or take two steps away from your monitor—you should still be able to tell relationships such as major and minor divisions and what paragraphs go together. If it turns into a big gray block or a string of indistinguishable smaller blocks instead, that's bad.

Classifying Your Documents

To know whether a document hits its mark, you must understand what the different marks look like and which one you're aiming for.

When Is a Manual Not a Guide?

Okay, you caught us, it's kind of a trick question—but not really. You see, it all depends on how you define (for your company or organization) the types of documents you create—and what to call them. If within your company you all agree that a document containing a certain type of information must always be called a manual, a manual it is. In that case, a manual is not a guide, because it's a manual!

Don't feel as if you have to reinvent the wheel, though. You can follow what others have done to address this very issue. For example, *Microsoft Manual of Style for Technical Publications, Second Edition* lists two pages of their standard names for types of books along with the intended audience and the type of information readers expect to find in it. (Look under "Titles of Books.")

You needn't feel bound to use these titles and profiles, but they're a good place to start. Just for the record, as far as we're concerned, "manual" and "guide" are two names for exactly the same thing. We admit there are arguable nuances of meaning between the two terms—and we've heard writers argue about them for hours without reaching consensus. Personally, we think there are better ways to spend our time.

Coffee Break

It's amazing how quickly words can change to reflect changes in society. As recently as the early 1900s, the term "broadcast" meant simply sowing seeds over a wide area. What a leap to today's "webcast" and "netcast." And yet, these new words still maintain their connection to their humble beginnings—as a human hand formerly cast seeds over fertile ground, it is still human hands on keyboards that disseminate and gather ideas across the planet.

The point is that you need to choose what you're going to call the different types of documents you'll write, then stick to those terms faithfully. That way neither you nor your readers will waste time wondering if there's a difference between a User Guide and a User Manual (or a User Handbook, for that matter, but let's not go there …).

Defining the Document Content

Once you've decided what to call the various documents, the next step is defining what type(s) of information each document category will contain—a document profile. This not only lets readers know what to expect if they pick up a guide, manual, handbook, or reference, it also helps prevent "document drift" over time.

Document drift is what happens when unplanned bits and pieces get added to a document as a product is changed and enhanced—like an old farmhouse that's had a room added here, a dormer added there, and a mud porch tacked on at the back. Pretty soon nobody is quite sure exactly how it is supposed to look. This loss of clarity and focus can only cause problems.

We recommend that you make a conscious decision to avoid overlapping information from one document to another. Tempting as it might be to help your user (usability again!) by presenting all the information right there, repeating information from one manual to another can be a disaster.

The main reason for not doing this is the difficulty of maintenance. When updating documents, it is a sure thing that you'll update information in one place but not another. It's pretty embarrassing to discover after many releases that one of your documents contains outdated information that is correct in another volume of the document set!

The other problem is consistency. With various writers working on documents, one person will tweak the writing in a manual without realizing that it is a duplicate of information found elsewhere. After a few rounds of this, we wind up with two same-but-different passages in related documents. And the user, as explained before, is confused, wondering if the slight differences have significance.

Instead, make sure you have a thorough index in each document (more about that in Chapter 14, "Indexing") and cross-references everywhere. Cross-referencing means that when a user needs to know where related information is, you tell the user where to look. If the document is online, the cross-reference is in the form of a *hyperlink,* so that clicking on the link jumps the user to the referenced document. Your word processing or desktop publishing program is sure to have a cross-reference feature. Find out how to use it—it's important.

With a document profile to refer to, writers can make good decisions about where new information needs to go. Are there new steps in installing a product? Put them in the Installation Guide. New ways to customize reports? That goes in the User Handbook. Do users need practice to master a new feature? Add that to the Tutorial.

A profile statement for each document type needn't be lengthy—take another look at the simple table format used in *Microsoft Manual of Style for Technical Publications.* The profile for each document just needs to be explicit and to have minimum overlap with other documents.

Tech Talk

A **hyperlink** is a link in an HTML document that jumps the user to somewhere. Hyperlinks usually are underlined or shown in a different color from the surrounding text. They can be pictures as well as words. **Hypertext** is text that has hyperlinks.

The Least You Need to Know

➤ Good documents are accurate and complete.

➤ Remember your audience. Think "usability."

➤ You have to know what a document should look like to decide if it succeeds.

➤ Avoid repeating information from one document to another. Use cross-references to send the user to the right place.

➤ It's important to define your company's documents, create document profiles, and follow them faithfully.

Part 2

Tech Writers, Start Your Engines ...

When you get hired for the first time as a technical writer, your employer expects you to know and be able to do certain things. Do you know what these are?

When you finish reading this part, you will. These are the things you need to know already when you walk in the door that first day—like what's involved in updating an existing document, how to learn about a product that's completely new to you, and how to focus a document's content to suit your audience.

Perhaps the most important thing you will learn from this part is the five steps of document development that are at the heart of all technical writing. (Yes, there really are only five.)

Chapter 6

Five Steps to Creating a Technical Document

Although different companies use different procedures, the steps for creating documentation basically are the same everywhere. Certain activities are necessary whether you are taking over a document written by someone else, or are writing a document from scratch—and each takes time. You might spend more or less time and effort on any one of the steps, but you must spend *some* time on each of them.

In this chapter we also give you an example of a plan that will help you through the process of creating a specific document. The document plan lists all your projected milestones, but you can't even meet the milestones unless you progress through the five steps.

Step by Step

Although we have given an estimate of the percentage of your time spent in each of these areas, *Your Mileage May Vary* (or *YMMV*). The time will vary according to the job you're doing and the number and depth of changes the product goes through.

All the parts of this process overlap with each other, however. Gathering information is a constant activity; writing might continue while the document is being reviewed and tested. Audience analysis continues as the audience changes or grows. And reviews happen over and over and over ....

1. Gather Information

Gathering information means everything from learning the product to interviewing developers. This is a research phase and it involves several stages.

Gathering information takes about 50 percent of your document creation time.

➤ **Read** product literature and specifications as well as any supporting documentation. We cover this stage in this chapter and touch on it again in Chapter 8, "Learning Your Topic."

➤ **Learn** by using the product so you can understand what it does and how.

➤ **Lay your groundwork** by conducting audience analysis and finding out what the document requirements are. Chapter 9, "It's All About Audience," discusses this phase.

➤ **Interview** developers and ask them for information about the product. Chapter 10, "Gathering Information," helps you get information from the *SMEs* (*Subject Matter Experts*).

➤ **Listen** by attending all the meetings you can. Often the developers forget to invite tech writers to the product development or other meetings. Let it be your responsibility to find out what meetings exist and ask to be invited.

Tech Talk

SME: Pronounced *smee*, this is a common tech writing acronym for **Subject Matter Expert.** A SME can be anyone from a product manager to someone working in shipping—it all depends on the subject matter expertise you need. Referring to "a SME" is also a shorthand way to convey the idea of "whoever it is that has this information" when you aren't sure exactly who that is.

It's just a humorous coincidence that this acronym sounds the same as the name of Captain Hook's assistant pirate in *Peter Pan*. (Honest, we didn't make that up.)

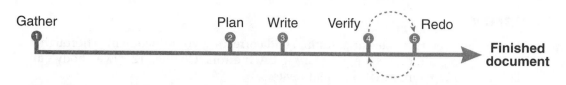

Gather   Plan   Write   Verify   Redo   **Finished document**

Knowing how much time to spend in each of the five steps is the key to creating an effective technical document.

(Drawing by Aaron Lyon)

2. Plan

Once you've gathered all the information you need and you have some understanding of the topic, it's time to plan and organize your information.

Planning takes about 10 percent of your document creation time.

> ➤ **Schedule** your work and make sure you can meet your milestones and match them to those of product delivery.

> ➤ **Understand** what your research means. If you have questions, now is the time to seek clarification. Only when you understand the material you have compiled can you write an effective document.

> ➤ **Collect** information that already exists to include in your document.

> ➤ **Organize** your material. Decide what the contents of your document are and organize those contents into the order you want.

Pros Know

"First and foremost a technical writer must be an advocate for the user to ensure that their view of the world is represented in the structure of the material. Everything follows from that. If the user lives in a universe where the sky is pink, it should look pink in every illustration, and be referred to as pink in every reference."

—Meryl Natchez, CEO, TechProse, www.techprose.com

3. Write

It can't be put off any longer! Now is the time to write the text. Time spent up front in planning and organization helps to cut down the actual writing time.

Writing will take about 20 percent of your document creation time.

> ➤ Draw up an outline to guide you, even if it changes as you write.

> ➤ Don't worry about how rough your first draft is; the important thing is to get something down on paper, and go from there.

> ➤ Be sure to go back and edit rough drafts at least a little, even if it's only to cut out extraneous information; your first draft is never your final draft.

4. Verify

Verification refers to checking, testing, and correcting your document. Thorough reviews by you and others are necessary to catch errors. Chapter 12, "Everybody's an Editor," contains information about reviews.

Verifying takes about 10 percent of your document creation time.

➤ **Check** your document by proofreading it to make sure it is correct to the best of your knowledge.

➤ **Review** your document by distributing copies to the SMEs and asking them to read it and comment on it.

➤ **Test** your document by going through the procedures to ensure the actual product matches what you wrote.

5. Redo

Even Jack Kerouac rewrote his work once in a while. No matter how in love you are with what you write, you will have to rewrite and edit your work. Chapter 15, "Making the Final Laps," contains tips for that.

Redoing takes about 10 percent of your document creation time.

➤ **Correct** your document by fixing any errors.

➤ **Clarify** your document by making it easier for the user to find and understand information.

➤ **Rewrite** your document to improve it and to add missing information.

➤ **Retest** the document by returning to Step 4, Verify. Repeat this action as many times as it takes to get it right, or until you run out of time.

That's the bulk of the document life cycle. Your content is developed and the next step will be to hand it off to production.

Creating a Document Plan

Before we create one, let's define what a document plan is and what it does: It is a document that spells out how you intend to get from where you are to where you want to be. The work is worth the effort because a solid document plan supports a writing project in many ways.

One of the primary functions of a document plan is to make sure that management, customers, and writers all have the same vision for the document and the same expectations about the process and the schedule. A manager usually creates an overall plan for all documents to support an entire product or family of products; an individual writer might work with only one document plan for each document he or she produces.

DOCUMENT PLAN

Title: *E-Buy&Sell 2.0 User's Guide*

Category: User's guide, release notes, online help for release 2.0

Date due: 6/29/2001

Author: Kim Reiter

Purpose: User's guide for beginning and intermediate user at home and on the Web; include installation info. Explain menu items and dialog boxes. Online help has to step through transactions and give examples of how to fill in each field as well.

Page count: UG: 40 pp. Release notes: 6 pp. Online help: 60 pp.

Contents:

What's new in this release

1. Introduction

2. Installation

3. E-Buy&Sell User Interface

4. Using E-Buy&Sell

5. Appendixes

6. Index

Milestones:

First draft: 6/1/2000

Second draft: 6/22/2000

A document plan helps all members of the team have the same expectations by specifying the goals, content, and milestones of a document at the beginning of the project.

A document plan is an outline of what you intend to do on a single document. Start with the basics:

1. Document title. (If you're not sure of the exact wording, give it a "working title.")

2. Document the category, product name, and product version. (Is it a Web site? tutorial? user guide? API?)

3. Delivery date. (This means finished documents.)

4. Your name. (The easiest part!)

5. Description of the document to be updated or written, including what its purpose is and roughly what kind of information it needs to contain.

6. Estimated number of pages. (This is a tough one—use existing documents of the same type.)

7. Approximate chapter titles. (This doesn't have to be as detailed as an outline would be.)

8. Your milestones and the delivery dates for those milestones.

Milestones

Originally, milestones were actual stones set into the ground beside primitive roads. They named the nearest city or town and stated the number of miles remaining to reach it. Milestones in tech writing perform a similar function. They are the interim goals you must reach and pass on the way to your final destination, the finished document.

Here is a list of very basic milestones you should plan for, shown in the order you'll reach them:

1. *First draft*
2. Review
3. *Second draft*
4. Review
5. Final version distributed to reviewers
6. *Camera-ready* version completed and submitted to the printer (or production department)
7. Printed and electronic documents in the hands of customers

On your document plan, you might want to fill in what you plan to provide or what's involved for each of these, so the reviewers and management know what your expectations are—and you do, too. For example, you need reviewers to return their feedback to you promptly so you can meet the milestone of a revised draft.

It is a good idea to add these details to the milestone list. The reviewers need to know you expect them to set aside enough time to properly review what you give them, for example:

5. Final version distributed to reviewers

6. Final comments due back from reviewers

7. Comments incorporated

8. Final document ready

9. Camera-ready version completed and submitted to printer

10. Printed and electronic documents in hands of customers

Once you have a document plan and know your schedule, it's time to get to work. The next chapter, "Getting Started," helps you begin your first assignment.

Tech Talk

Camera-ready is a term that comes from printing production. A camera shoots each page or set of pages as the first step in making these plates. Now, camera-ready copy may never go to a camera, but go straight to a DocuTech printer from a PostScript file. Nonetheless, the vocabulary lives on. Camera-ready art or copy is, literally, picture-perfect and ready for production.

Pros Know

Even though a first draft is incomplete and rough in some ways, it's still a good idea to proofread, spell-check, and run a table of contents and index on it before you send it out for review. Do what you can to make it as complete as it can be. Reviewers worry when they see obvious errors and typos and may assume the content is incorrect, too.

The Least You Need to Know

➤ Follow the five steps of document creation.

➤ Make a detailed document plan to understand and agree on your goals.

➤ Break a project into milestones to make the parts manageable.

➤ Don't worry about perfection in your first draft, but do go back and edit before the review.

Chapter 7

Getting Started

In This Chapter

➤ Taking over someone else's document

➤ Knowing when to rewrite and when to leave it alone

➤ Seven ways to make the developers think you're an expert

"Getting Started" is a common section title in beginners' software books, and that title is relevant for this chapter. Although we devote a lot of this book to the process of writing brand-new documents where none existed before, your first forays into technical writing probably won't take you down that long and winding road. In your first job, you're more likely to be given proofreading or editing work on something that already exists.

Once you've got your feet on the ground with that, you'll likely move up to ownership of a document that needs updating, maybe with extensive revisions. When you've finished that, you'll have been around the block more than once and will be ready for the trek into writing a full, new document from scratch. In the high-tech world, that chance might come sooner than you think!

In this chapter we'll guide you through updating and maintaining documents you didn't write so you can avoid common pitfalls and wrong turns. Even if your ultimate destination is to work from home as a consultant, there's no doubt that sooner or later you'll walk the update road. So think of your journey of a thousand tech writing miles as beginning with this first step.

Your First Assignment: Taking Over a Document

Although we've known some companies in which brand-new, inexperienced writers were expected to start writing from scratch on their first day at the job, you'll be relieved to hear that this situation is rare. A manager wants to help you learn the product and become familiar with the department style, and usually will gradually work you up to the tough stuff. Generally this means starting you out on proofreading, copy editing, or updating a document someone else wrote.

You might be surprised at being asked to edit before you've written, but this progression actually makes sense. It's always easier to revise and improve what someone else has written than to create original text yourself. And because you won't be asked to make judgments about the content, you can apply the writing skills you already have while you're still learning about the general look and feel of the company documents.

You're also exposed to a tremendous amount of new information as you're proofreading or editing, so you're really doing double duty—working productively and acquiring product knowledge at the same time.

Adopting a Document

Rather than tell you about proofreading and editing here (these are covered in Chapter 15, "Making the Final Laps"), we want to walk you through what to do when your manager announces you are responsible for "updating" or "taking over" an existing document. What exactly does that mean?

Usually it means you've adopted a document, just like bringing home a stray animal from the pound, and it has now become your companion on the road. No matter what its history or how it came to be the way it is, its welfare in the present and near future now depends on you. Your immediate task is to make sure that new information is added appropriately (such as

---

Pros Know

As a professional, it's important to work like a duck: Look cool and calm on the surface even when you're paddling like crazy underneath. Experience is the best teacher, but try to think ahead and anticipate what you'll do if something doesn't go exactly as planned. Alert your boss promptly (and calmly) if you sense "a change in the wind."

Dodging Bullets

Taking over someone else's work requires tact. It's very bad form to complain about how badly written a document is. You don't know the previous writer's circumstances—because of limited time and resources, it actually might be a glowing success. That writer also might still be around in your department. You might even discover—too late—that the writer now is your manager! So if you can't say something nice ...

when a new software release is in the works) along with fixing known mistakes or omissions. You also might have a free rein to rewrite parts that need it and certainly are free to rewrite parts that don't conform to the current department style.

"Ownership" of a document can be pretty fluid in the computer industry. You might be asked to update a document strictly by adding new information, while the ownership of the document stays with another writer who is still responsible for addressing errors, omissions, and style compliance. Or you might be asked to "take over" a particular document during the intense later stages of revising for a new software release, but give it back to the original writer once the deadline is met.

Do your best while the manual is in your care. If you're not sure what to do, ask your manager to clarify exactly what work you should and shouldn't do to a document assigned to you.

Also remember that just as a document was handed to you, the care and feeding of a document you've had ownership of might be shifted to someone else. So don't get too attached. A document can change so much over several revisions or software release cycles that you'll hardly recognize any part of it. And that's as it should be. After all, the document is about the product, not about you or what you did to it.

Pros Know

"Perhaps one of the toughest issues for a new tech writer is to determine what your job actually is. Ask direct questions about what is expected in your job performance, and if possible, get it in writing. This not only protects you, but makes your manager think about task assumptions and the required expertise ...."

—Walden Miller, Director of Engineering Services, Vidiom Systems, www.vidiom.com

Maintenance and Tune-Ups

Not all tech writing can be a daring adventure, scaling the peaks of new technology. Some of it is plain old plodding, just putting one foot in front of the other. Maintaining documents, although important, definitely falls into the latter category.

Out with the Old, In with the New

Users depend on the documentation to be accurate and to reflect the current state of your company's product, service, or whatever. That product goes through new releases; in the computer industry that can be quarterly, with smaller updates and fixes happening at even more frequent intervals. So it's imperative that documents are periodically combed through and brought up to date. There's no denying that this necessarily methodical process lacks excitement. In fact, it's one that people sometimes conveniently "forget" about—to their detriment. But sometimes obsolete information is worse than none at all.

Tech Talk

Change requests are official requests, usually submitted through a formal internal system or process, for error (or "bug") correction in software or documentation. These requests typically are entered into an automated system upon receipt and then ranked in order of either their importance to the product or the ease with which they can be corrected.

Dodging Bullets

Don't start something you can't finish. Keep the document as a whole in mind when contemplating significant changes to a single part, chapter, or section. Sudden differences in appearance, style, and tone are jarring to the reader and ultimately reduce the document's effectiveness. If you can't make a desired change evenly throughout the document, don't do it anywhere.

Be guided by the reason you're updating the document in the first place. Was it something created on an incredibly short deadline as a stopgap that now can be "done right"? In this case, substantial rewriting might be exactly what's needed. But if the document is being revised to reflect a new software release, now probably is not the time to begin a serious rewrite. Find out.

Your first priorities are adding new information, making corrections, and filling in missing parts and pieces. Often the documentation team collects *change requests* and saves them for a product release—make sure all these are addressed as well. Then, if you still have time, see what rewriting you can do evenly across the document that will make the greatest improvement for the least time and effort.

Knowing What Needs Fixing

Changes and corrections aren't always collected for you on a tidy list—frequently you'll have to ferret them out by attending development meetings and talking with engineers, product managers, developers, and sales reps. And not every change someone else considers a good idea has to show up in the next document version; as you learn more, you'll be able to use your judgment about which ones to include.

It doesn't help matters that revisions can take nearly as long as writing a new manual. You might think a document needs just a little change here and a little addition there, but even a routine update must address all of the following:

➤ Update document ID numbers, release numbers, and release or version dates in the documentation.

➤ Find out if any customers or colleagues had change requests or feedback that needs to be addressed.

➤ Update the template used to create the document if necessary.

➤ Add new features and functions or other new information. (Do you know what's changed in the new release? Find out.)

➤ Correct known errors. (Does someone in the department know what they are? Who?)

➤ Add information known to be missing from the current version. (Again, find out who can tell you what's missing.)

➤ Check all screen captures, diagrams, tables, and so forth, to make sure they still reflect the current state of the product. (Look carefully—screen captures in particular tend to change with each release but differences aren't always obvious.)

➤ Clarify and rewrite as necessary. (This is where your own judgment comes in.)

➤ Update the index entries where the text was changed.

Pros Know

If it ain't broke, don't fix it. Wading into a major rewrite simply because you think your writing style will make it better is almost never a good enough reason. It may even be something your manager considers a waste of time. Discuss big overhauls with him or her before you start—it won't make you look good to spend valuable time on a job that didn't really need to be done in the first place.

How Much Rewriting Should You Do?

Although the wordsmith in you might be itching to polish, revise, or completely rewrite parts of the document or even the whole thing, be cautious about how much of this you do. It's easy to spend too much time rewriting and not leave enough time for more material changes. If a document's information is accurate and complete, minor text polishing to improve clarity probably is enough.

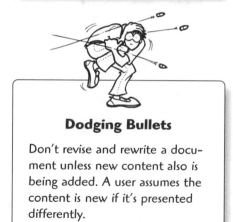

Dodging Bullets

Don't revise and rewrite a document unless new content also is being added. A user assumes the content is new if it's presented differently.

Even if rewriting *would* improve the document, it's still not necessarily a good thing. A user accustomed to seeing information laid out in a particular way won't appreciate having to search for it in the next revision. In addition to rewording text, a revising writer tends to move things around and chunk the content differently. A reader who is familiar with the last version of the document is going to be annoyed at having to search for information he or she was familiar with. The user might miss important information or skip over essential steps.

Where to Begin—A Plan of Attack

By now you have an idea of where you want to go in updating a document, but you still might not be sure exactly what to do first when given an assignment. Here are some guidelines to get you pointed in the right direction:

Dodging Bullets

Be careful what you call "yours" when offering samples at a job interview. Only if you wrote from scratch can you claim full authorship. If a sample was originated by someone else, make that clear even if you revised it extensively. Point out the new parts you wrote yourself. And don't think nobody will find out—we know managers who have interviewed candidates claiming authorship of documents the managers themselves originated!

1. When you take ownership of your first document, read it, or at least skim it, as soon as possible. Your manager and the developers will expect you to be familiar with the document's contents, and you should be able to answer questions that come up about what is and isn't in it. Assume that this document will be in your care indefinitely. Treating it as if you will own it permanently means you'll read it, understand it, and become an expert on it—a good thing whether you keep working on it or not.

2. Find out which Subject Matter Experts (SMEs) are the resources for this document's content and ask to be introduced to them, or go introduce yourself. Set up a time to go over your document with each of them to hear about what's new, and ask if there's a product description or change specification you can read before the meetings (then be sure to do so).

3. If the document's last writer is still available to you, talk to this writer about things that need to be changed. Writers rarely have enough time to do everything they want to do, so the writer might already have a list of areas that need work.

4. When you feel that you've looked at enough of the manual to have an idea of what the topic is about, ask for a demo of the product. Ask your manager or project manager for the name of the best person to give you this demo, then go schedule it as soon as possible.

5. Use the product to see how its reality compares to what's in the document. Be careful about what product version you use—make sure it's the one that matches the release you are writing about. Many companies simultaneously update and maintain documentation for several versions of the same software for customers who don't want to upgrade. If you're updating online help for an earlier product release that your company still maintains and enhances, looking at the latest and greatest version of the product won't help you at all. Likewise, if

you're doing a major update to reflect a totally revamped product, working with the old version will be only of limited value, if any at all. Software releases have version numbers. Make sure the version number you use matches the version number you're supposed to write about.

6. Ask the manager who assigned you this document if any change requests have been opened for it, or if he or she has heard of any comments or complaints about the document such as from sales reps or help desk technicians. Collect these comments, print them out, and read them even if you don't understand them yet. Go talk to people who have complained—even if they've done it only conversationally.

7. By the time you meet with the SMEs, you should have a fairly good background in the content you'll be working with. Although it's certain that you won't know everything, make it your responsibility to know enough that you can ask intelligent questions and understand the answers. There's no need to pretend to understand something when you don't. They'll understand if you're new and usually will be happy to show you their "baby." Make it a priority to learn what you can before going to meet with the people who will provide you with information. Have your questions ready and go over the change requests with them if you have questions. Read Chapter 8, "Learning Your Topic," for help with learning about your topic.

Also understand that it's the developers' jobs to explain things and answer questions, but don't expect them to write the information down and give it to you verbatim. Providing information is their job, but capturing it and making sense of it is yours.

Coffee Break

Where did the term "bug" come from? Nobody's sure, but it was already in use in 1947 when a moth got between two electrical contacts and shorted them in the huge Harvard Mark II computer developed by IBM-backed Howard Aiken and his team. A technician's log entry noted that it was the "first actual case of a bug being found" and he taped the proof—the moth itself—into the logbook. (Now, that's what *we* call documenting!)

Dodging Bullets

A sure way to *not* make a good impression is to go to the head developer on a project and ask him or her to tell you everything there is to know about the product you're documenting. If you tried your hardest, but still couldn't learn about the product, ask a lower-ranking developer for a demo. Then ask questions as you watch.

If There's No Time ...

It's not unusual for your first assignment to come during a real crunch (your manager might even have hired you because he or she saw this crunch on the horizon weeks or months ago) and you might feel panicked. There's no time to learn about the product, and your boss expects you to do a job that, from where you stand, looks like jumping the Grand Canyon without even getting a running start.

Estimating your time and meeting deadlines is a real trouble spot for a new writer. (Hey, it's difficult for a lot of old-timers, too.) At the beginning of an assignment, you probably aren't even thinking about the deadline, even though it might be a mere week away. Well, that's when it's important to think about your deadline.

Break Down Your Schedule

As soon as you know what date you need to finish the job, break down your schedule into sections. If you have several weeks to finish the job, estimate how much you need to have finished each week, then break it down into days. If the assignment is merely days long, calculate how much you need to do each hour, if necessary.

Check Your Progress

Check your progress regularly. If you are working too slowly, talk to your manager as soon as you believe you might not be able to finish, and ask where you can cut corners or how you can get help. If you are working too quickly, don't assume that means you can take time out to do something else. Until you are very experienced at estimating your time, try to do as much as you can up front on each assignment.

No Matter What Happens, Stay Calm

And if no matter which way you look, it seems you've been given an impossible assignment that no human could finish in the allotted time? Slow down. Take a deep breath. Then take another one. Suppress that urgent desire to bolt. Go through the seven steps in the previous checklist, even if it seems you're wasting time—you're not. You'll be able to start making strides in the right direction instead of running around in circles as if you haven't got a clue.

After you've been on the project through a few product life cycles, you might just realize that the first assignment wasn't much of a time crunch at all!

The Least You Need to Know

➤ Be ready to assume full responsibility for the document; it probably will be yours for a good long time.

➤ Prioritize changes that need to be made to your document—address new information first, fix errors and omissions second, and leave wordsmithing for last.

➤ Learn a document's contents and history and what's needed for its immediate future before you plan a major rewrite.

➤ Educate yourself before you visit the developers so they can be confident about your knowledge and abilities.

Chapter 8

Learning Your Topic

In This Chapter

➤ How and where to start

➤ Why product knowledge is important

➤ How to ask without looking foolish

When you begin a job or are presented with a new product, it can seem impossible to learn everything you need to know. There might be physical obstacles—you might not have the appropriate equipment to use the product, or (at small or disorganized companies) it might be days or weeks before you even have a computer on your desk. If you're new on the job, you don't know anybody yet and the thousand or so names you've heard have flown out of your head. And what's the name of that software you're supposed to document, again?

This can be a challenging situation. But don't let it stop you from starting to learn as much as you can about the product you'll be writing about. This chapter not only explains how you can do that, but it also shows how you can study the product with an eye toward putting that first-time-user experience into the customer documentation.

The Importance of Product Knowledge

We've alluded to this many times, but the most important thing you can do to write good documentation is to learn about the product. Nobody expects you to be able to divine a product's inner workings and plumb the depths of its soul without help or resource, but we think it's essential that a writer know as much as possible about the product at hand. Why? There are lots of reasons, all of them good ones:

➤ You will gain the respect of the developers and be able to talk to them with some intelligence (this is a biggie).

➤ It enables you to determine whether information given to you is correct (this is another biggie).

➤ It enables you to write much faster than if you have to grope your way through the information.

➤ You will gain an overall picture so you know whether the documentation is complete.

➤ It prevents you from asking the same questions over and over (which goes back to the first reason).

Getting to Square One

Any or all of the following items may take several days or even a week or more to get in place, so ask for them or confirm that you'll have them *immediately* when you learn there's a product for you to write about.

Learning the product might not be as simple, though, as walking over and sitting down at an open terminal or computer and logging on. You might need to have someone else set up many of the following before you can even begin:

➤ A computer where you can run or access the product when you need to without interfering with someone else's work, or having someone else interfere with yours. (This might be the computer on your desk, or it might not. Hope that it is.)

➤ A user name or login name (and usually a password) within the product or the product's system.

➤ The same user privileges your audience will have.

➤ Dummy directory structures or database tables you can use that won't affect test data or the product development environment.

➤ Access to the latest and greatest version or release or model of the product. A demonstration version won't do. You need access to the real thing, even if—or especially if—it changes every day.

Product Ins and Outs

You've got product access, you've got a login and password, you've got a cup of coffee and a pad of paper and a couple of hours. You're ready to dig in. Now what?

There are a number of different ways to approach learning about a product; all of them work. The important thing is to pick the way that works best for you.

➤ A **methodical approach** begins at the first screen and works sequentially through every screen and every option, in order. The benefit of this approach is that you are sure you've seen everything at least once. The drawback is that you might not get a sense of how the product accommodates different user task flows; for example, you might not see how editing existing database entries differs from adding a new entry.

➤ A **task-based approach** begins with a number of tasks the user might want to do, stepping through the screens needed to do each task. The benefit of this approach is that not only do you get a good sense of how a user experiences the product in real life, you get a leg up in understanding how the product fits into the user's work day and why he or she is using it in the first place. The drawback is that unless you can perform every task the user might do, you can't know you've experienced the whole product or put it through all its paces.

➤ A **"jump in and swim"** approach begins on the first screen and could go anywhere from there. To an observer this approach might appear to consist of random wandering among the product's screens and options, but for some people this is the most natural way to gather the individual pieces and start fitting them together. Potentially, this approach has the drawbacks of both the methodical approach and the task-based approach—but if this is *your* style of learning, you know it also can offer the benefit of deeper insight and understanding than observers (whose styles are different) might guess.

Perhaps the best approach is a thoughtful combination of more than one of these. (We recommend sticking with the methodical and task-based approaches unless you already know your style of learning really is "jump in and swim.")

Leave a Paper Trail

Remember as you first use the program or product that this is your first and last chance to experience it from the perspective of the totally naïve user. Take advantage of that: Be aware of what you don't know and how you search for information, and make notes of the things that are difficult or not intuitive. In fact you should write everything down—you'll need it later when you become more familiar with the

product and forget how it was to use it for the first time. It's easy to get carried away exploring the product, but try to rein in your enthusiasm a bit—in your eagerness to learn, you could be missing your chance to put yourself in the user's place.

Capture Your "Newbie" Experience

The other (and probably hardest) thing to remember is to make yourself go slowly. This is not the time to show off how quickly you can master something new. Take time to focus and think, and make notes that will be meaningful to you (or someone else) later on.

The notes you make during this first exploration session will be critical in developing your document. Take time to make them clear and thorough. Use bullets to capture related points quickly—a small tick beside a phrase will do—and be sure to number steps or paragraphs when sequence is important.

Use whole words, too, for product features and functions, especially at first; you'll slip into a natural "shorthand" notation quickly, and it's essential to have a reference point where the actual terms your notations refer to are spelled out clearly.

Write down even things you are sure you'll remember. This helps in several ways: You don't have to rely on your memory (which, if you're like everyone else, can have a way of playing tricks), and you have something tangible to refer to and share with others—who might not trust your memory, even if you do.

Tech Talk

Bio-break, although not a true technical term, is one you'll be glad you know. Use it when you need to ask tactfully for a break in a meeting or interview to "answer the call of nature" or attend to other biological functions such as the need for a drink or snack ("How about a bio-break, everyone?"). This is a welcome reminder that our co-workers are not machines, and neither are we.

Label Notes Clearly

If you're going through a product screen by screen, you might want to start a new page of notes for each screen. Be sure to write down the name of the screen (and number, if it has one) on the first page. Carry this screen name over to subsequent notes pages about that screen, so if your pages get shuffled you'll be able to straighten them out again. Be sure to number your notes pages, too, and include the date. The date can be an important clue later if there's any question what version of the product you were using.

Keep Notes Based on User Tasks

If you're focusing on a collection of user tasks, make a list of all the tasks you want to explore on a separate sheet of paper, then assign letters to each task. Begin each task's notes on a new sheet of paper, with the task name and letter at the top. Then, you can just use the single-letter notation on subsequent notes pages about that task. Remember to number and date all your notes pages.

Ask and Ye Shall Learn

Knowing document design and development is a big plus for a tech writer, but all the document design in the world won't help if you don't know which information in the document is wrong or what needs to be emphasized. We'll discuss some ways to sharpen your product knowledge in the following sections.

Pros Know

It's a good idea to use a stapler to join related pages of notes instead of a paperclip. Paperclips have a quirky habit of picking up unrelated pages or letting the back page of a group slip away, and they can be popped off accidentally. Staple holes in your originals can also distinguish them from photocopies, when that's important.

Become an Expert in the Field

Become educated in the field you're writing about. If you're writing about UNIX administration, study UNIX. If you're writing a user's guide for a cell phone, take a course in wireless telecommunications basics. Having expertise in the entire area will help you learn the specifics you need to know about your product.

Buy a book on the general subject at your favorite bookstore and read it to get an overview of what drives your product. Better yet, get one recommended in the trade press for your industry.

Learn the Specifics

Collect all the information you can find about your product. Find out if anything was written by developers or product planners before the product was created. (Not every company creates programming or development specifications, but if yours does, get copies—you will be more than happy to have them.)

Marketing is a big source of information. They often have a clear idea of what the product should do before it even exists. Get a copy of all the marketing material you can find that relates to the product. Ask the people who developed the marketing materials where they got their information.

Visit your company's Web site and read all the marketing materials about the product that you can find. You might even want to print out selected pages and keep them for offline reference. This might seem like an obvious way to learn about a product—and it is—but that doesn't lessen its value. And, if the marketing people have done their jobs, you can learn a lot about your product's users from the marketing approach.

Play Around

Start playing with the software as soon as you can. If you're one of those people who can't figure out how to start the application, ask! People are glad to help get you started, as long as you don't expect them to stick around and hold your hand.

Ask your boss or co-worker who is the best person in the company to show you how to use the software or walk you through a demonstration. Before the demo, ask for a verbal or written overview of the product so you can better understand what you're seeing. If your company's product is hardware, ask to see one with the cover off, and ask what the components are and what they do.

After the demo, start using the product yourself. Investigate every menu command, every dialog box. Check for new menus when you go to different parts of the program. Hit keys randomly in various combinations and see what happens. If it's command-line software, ask where you can get a list of commands. For hardware, be sure to get a list of "don'ts" along with your product orientation.

This experience will raise questions that will be valuable input to your documentation. How does the user choose between options offered on a screen? Where does the information that goes in an entry field come from? When the user has done everything correctly, what has he or she achieved?

Pros Know

"If you truly understand the topic, writing documentation is a snap. Once you hold knowledge in your head, manipulating, exploiting, leveraging, and condensing that information into bite-sized chunks is a breeze. You just have to be able to handle the wrist-aching monotony of pecking away on a computer all day."

—Andrew Plato, President/Principal Consultant, Anitian Consulting, http://www.anitian.com

Meet the People Who Can Help You

Find out who the product's project leader or program manager is and ask to be introduced to the people responsible for the different product aspects. Go to each of them and ask for a five-minute demo of their part of the product. They'll get to know you, and you'll learn something valuable from each of them that you can follow up on later.

Ask to Share the Wealth

Ask people you meet if they have any notes or documents they'd be willing to share with you or let you copy, or if they can recommend anything for you to read. You will collect a lot of information this way and become less of a stranger to your colleagues!

Also go to the developers' meetings. In fact, at first, attend every meeting you can. Take notes. They probably won't mean anything to you for a while, but later you'll be able to understand them.

Pros Know

When you go to meetings for the first time, don't feel bad if you don't understand anything. The acronyms and corporate buzz–words will make it sound as if you're listening to a foreign language, and until you understand the product, the technical talk will be equally incomprehensible. Just sit and listen. No one will expect you to contribute anything for a while. After a few sessions, you'll be an old pro.

Learn About Related or Allied Products

If your company makes a well-known product with lots of existing documentation or interacts closely with one, read as many documents that relate to it as possible. For example, if your product runs on an Oracle database, there is a wealth of material about Oracle and SQL for you to read.

For software products, take a look at the code if you can. Buy a simple book about the language your product is written in and start learning it. You can learn a lot about how the program works by reading code, and this also gives you access to developer comments, which usually are written into the code.

Take Your Time

Last but not least, give yourself time! Don't try to do all this at once! As time passes you will know more, and by the second release of your document you will start to feel like an expert.

The Least You Need to Know

➤ Get the right equipment *right away* so you can start learning.

➤ Acquire an education in the field, not just in your specific product.

➤ Ask colleagues if they have notes or documents they would be willing to share with you.

➤ Go to meetings and take notes, even if you don't yet understand everything you hear.

➤ Don't hate yourself if you don't learn all there is to know about the product right away. It takes time.

Chapter 9

It's All About Audience

In This Chapter

➤ Discovering and defining your users

➤ Finding out what your users need to know

➤ How to interview potential users and where to meet them

➤ How to fake it if you can't meet the users

➤ How to write for a user who has more technical knowledge than you do

So now you've learned about your new job, made a plan for your document, and laid out your schedule. So you're ready to write—right? Well, not quite.

Before you lay pen to paper or fingers to keyboard, there's a secret that every good tech writer should know: *Know your audience*. Once you know that, everything will come easily. (Well, okay, that's too much to expect—but it will be lots easier.)

Journalism has the five Ws—the questions you must answer to write a good story— "Who, What, When, Where, Why." In technical writing, the all-important question is "Who?"

Who Will Read This Document?

Whenever you start working on a document, you first must find out who is the intended *end user*. Ideally, your boss or the product manager has already thought about this and can answer it for you, or at least give you a place to start. You'll find that getting an answer to that question will simplify your job tremendously.

Tech Talk

The **end user** is the intended user of your product or document, dif-ferentiated from the people who designed or created it.

In general your user will fall into one of only a few categories: novice user, power user, programmer and software developer, and system administrator. These categories are defined by how much the user already knows and what the user wants to do. Some documents serve more than one user category.

Novice User

The novice user has never seen this program or piece of hardware before or has never encountered the topic at hand. The novice needs examples that work.

You must assume that the novice knows nothing about your program, hardware, or topic. Often, a novice is simply using a program or piece of hardware to accomplish a business or personal task (for example, word processing or database updating at the office, or playing a game at home).

Novices need something short and simple with which they can get started quickly. They do not necessarily need or want full explanations because their immediate need is to get up and running; after that they might refer to other documents for additional information. The novice user, of course, might turn into the power user.

Examples of documents for the novice user include tutorials, introductions, user's guides, installation guides, and quick reference cards.

Power User

The power user already knows how to use the program or hardware and needs reference material and details about how to do more than just the basics. If you're writing about an industry topic, this person is already familiar with the topic's issues and terms.

The power user is an adept user who has been familiar with a product for some time, and who knows how to use it to perform work tasks. The power user category might include technical support representatives or customer service people—those whose job it is to help customers with problems or questions. It also might include presales engineers and postsales installers from your own company.

Like novice users, power users often will use the program or hardware to accomplish a business task—often complex, enterprise-critical ones. They need many of the same documents, but will use them in a different way.

For example, a novice might use a quick reference card to get started, whereas a power user wants the quick reference card as a refresher, so he or she doesn't have to keep the documents around. *Release notes* are important to the power user, because he or she knows the product and is interested in upgrading.

Examples of documents for the power user include user's guides, installation guides, quick-start cards, troubleshooting guides, reference guides, command and topic indexes, and release notes.

Programmer or Software Developer

The programmer might be customizing the program or using it to do development work, either within your own company or within a customer's company. This user is concerned with understanding how the program or hardware works rather than with using it to accomplish a business task.

Documents for this user include APIs and reference guides of all types.

System Administrator

The system administrator has the job of making a company's system or network run. The system administrator (or sys admin, for short) works with in-house programmers or software developers to optimize your company's product.

Documents for this user include system administrator guides, reference guides, and troubleshooting materials.

There are, of course, variations on the different types of users. You might be writing for specialized types of users such as trainers, sales representatives, marketing people, and technicians. In all cases, you will consider whether the user needs your documentation to help him or her perform a task (novice user or power user), or as information (programmer or salesperson).

Find out where your user fits in, and you will find out where your document fits in.

Quiz

Users come in all kinds of different packages, and each has a different need. Suppose you are

Tech Talk

Release notes are short, easily accessible documents (sometimes called README files) that accompany a specific release of a product. Release notes contain information that does not belong in or did not make it into the regular customer documentation.

Pros Know

It's crucial to be alert to what names your company uses for different documents. In some companies, a "user guide" is aimed at a user who's a system administrator. It could include everything from reference material to installation instructions; whereas in other companies a document with the same name might consist of only quick-start instructions, a tutorial, and a description of menu items. It's up to you to find out what a "user guide"—or other document—is at your company.

expected to write user documentation for scanner software. What type of document(s) would you produce for the following users?

1. The users are specialized graphic artists in the pre-press field who use the scanning software with expensive hardware. At least half the audience has used a previous version of this software.

2. The users are cashiers at a copy shop who have no personal interest in scanning but who are expected to do it quickly and accurately as part of their job.

3. The users are PC owners scanning at home on a hundred-dollar scanner.

4. The users are technicians whose job it is to repair scanners, some of which use this software.

5. The users are sales personnel and their customers. The customers work in the pre-press field but have never seen this software before.

Each user might require a completely different approach, and it will be your job to determine what the document type, content, and level of detail should be. See Chapter 16, "The Deliverables," for information about the types of documents you would deliver to these users.

You seldom will be asked to write a document that exists in a vacuum. Once you have a basic idea of who your user is, think about whether there exists other documentation already tailored for this user. Exploring existing documentation helps you learn ...

➤ What that user knows already (and therefore, what you can omit).

➤ What kind of tone is considered appropriate (friendly? formal?).

➤ What the user is expected to do with this product (the business tasks).

If no documentation exists for your user, you must decide how much information to include in the document you're about to create. There's a good chance you'll need to be comprehensive.

Matching the User with the Document Type

Audience analysis—even the briefest sort—can help you tremendously in getting started. Once you identify the intended audience for your documentation, you already have a fair idea of what content is required from these basic descriptions.

Coffee Break

Some people argue that "audience analysis" often is a waste of time. They envision the tech writer who spends weeks of precious writing time developing questionnaires or watching the customer use the documentation. That *would* be a time-waster, in our opinion.

But audience analysis can be tailored to the time you have available. If you are creating a new line of manuals for a company that has plenty of staff and is committed to usability, by all means, do all the audience analysis you can do—as long as it is productive.

If you only have a few days to churn out a big new document, perhaps all you'll be able to do is get a quick answer to your question, "Who's my audience?" as you fly past your boss.

Chapter 16 contains examples of the types of documents a user needs. Once you know who the user is, you can refer to the document types in Chapter 16; then you can make some pretty good guesses about …

➤ The content—what to include.

➤ The structure—how to organize it.

➤ The complexity of language—how to talk about it.

➤ The approach to take—descriptive versus task-oriented.

Fine-Tuning What You Know

The more you know about your user, the more and better you can write for him or her. Why? You can better identify the features that are important to that audience. (As you answer the questions in this section, you also learn whether the document set that accompanies your product is complete or whether there are pieces missing, and which ones.)

The questions in this section enable you to learn about who your user is, and what he or she does with the product. Let's look at these questions one by one.

What Is My User's Job Title?

Your user's job title gives you some clues to the type of material that user needs, as well as the user's likely level of experience. For example, "network administrator" tells you the user is a technical type who runs the system for multiple users and machines. "Bank teller," on the other hand, tells you it's important to know the typical business tasks bank tellers perform and how your product supports them.

What Is My User's Job Function?

Learning about your user's job function helps narrow the focus of the document. You might be writing a manual whose main users are the "professional services organization" of your firm, handling presales, installation, and customer service. That means your document might need to function not only as an installation and troubleshooting guide, but as a presales "leave-behind" piece, to be read by customers who haven't yet decided to buy. That's a tall order for one document!

Knowledge such as this can help you decide what information to include—or it can help you decide to produce two or three separate documents instead of one.

This is where profiling your user proves its value. For example, you might be asked to write something your boss calls a "user guide" for network management software. "Okay," you think to yourself, "my users probably will be system administrators and network managers." But then you start asking those essential questions—and discover that although some users indeed are system administrators and experienced network managers, others are administrative assistants and data entry workers who don't know the difference between a LAN and a WAN—and sometimes couldn't care less. You might need to back up and decide if a single document can really do the job.

What Tasks Is the User Performing with This Program?

What steps or tasks are involved in the job the user is doing? These tasks and steps move the user toward accomplishing a business goal—such as producing a finished image or making an e-commerce Web site handle orders smoothly. The more you learn about the types of jobs and tasks your users do, the more you know about what information needs to be included in writing about your company's product.

How Will the Reader Use the Document? And How Often?

Will the user refer to the document every day? Only occasionally? Will it perform a time-critical task such as teaching the user how to quickly master a product's basic features and functions? Or will it stand as a reference work that helps users find what they need quickly and easily? Knowing how someone uses a document helps you decide the kinds of information to include and how to organize it in a way that makes sense to the user.

How Does the Product Help the User?

Different products offer different kinds of value in a user's day. Some products are simply tools to get a basic job done, such as a word processor, a printer, or a scanner. Like the telephone or the kitchen sink, users tend to take these workhorses for granted—until something goes wrong.

Some products are designed to make the user's life easier, even though they might not be used every day—a money management program such as Quicken, for example, can be a huge relief to someone who hates balancing a checkbook every week or every month.

Some products become important professional tools, used specifically by members of a particular field. This could be something like accounting software, medical software, or an environmental program that tells workers who handle toxic industrial waste how to dress.

What Possible Problems Might the User Encounter?

Any predictable or known problems you can identify early on can become a basis for user guide content as well as fodder for troubleshooting questions and answers. Remember that if users have trouble with the product, the first place they'll turn for help is to the document you write. Your document can solve a problem and create a happy customer—or do just the opposite.

How Computer Literate Is the User?

This one can be tricky. If even some of your users are new to computer use, you must explain everything in short, clear steps. Never assume your user knows a term or concept that has become second nature to you and others with more experience. Remember, you had to learn it once, too!

But you can be in the opposite position, too—assigned to write for a technically savvy audience that puts you in over your head. We discuss that a little later in this chapter.

How Educated Is the User?

If you are writing for a product used in people's homes, you can assume that among your readership will be people who have little or no formal education beyond having learned to read. (If they can't read, you might be faced with a completely new and different challenge!)

If you are writing for professionals in any field, take some time to find out what level of education these people have. Read some trade magazines and publications, being alert to vocabulary, jargon, and the complexity of paragraphs and sentences. Find a Web site connected with the particular business and get a feel for the person it's designed for.

If you're writing for computer professionals or programmers, assume that your users will have at least a two- or four-year college degree or the equivalent in work experience.

No matter what the education level of your audience, however, you must always write clearly and as simply as the topic allows. Chapter 19, "Writing Clearly," discusses that more.

Is English the User's Native Language?

Some documentation is published in English and distributed worldwide to all countries, where the user is expected to read and understand English. This type of user can be assumed to have a fairly high level of education, but will not grasp English as fully as will a native speaker.

When your user is a non-native English speaker, make sure you avoid all Americanisms, colloquial terms, and confusing language. In some ways, the rules for non-native English readers should be no different from the rules for any reader!

Many companies now translate customer documents into other languages. For this product, you must consider the translator to be your user as well. Mistranslation of a key point or process can be a serious problem, and you don't want it to be because you were unclear in your terminology.

> **Pros Know**
>
> "My most humbling yet valuable experience as a technical communicator came the first time I saw what users were actually using instead of the manual that I was so proud of. After spending lots of time wondering what was wrong with them (the manual in question was one I'd produced using advice gleaned from books, articles, and conference presentations by the best and brightest in our business), it finally occurred to me that there might be something wrong with the manual."
>
> —George Hayhoe, President, George Hayhoe Associates, http://www.ghayhoe.com

Same Software, Different Users

Here's an example of user profiling and how you might approach it as a tech writer: Suppose you have just been hired to write documentation for Internet auctioning software. Let's call it E-Buy&Sell.

The Home User

Home users can subscribe to E-Buy&Sell from the Internet and engage in online auctions. Your manager asks you to write online help for a home user who is bidding one at a time on some items. The user needs instructions that outline the bidding process and describe what each button on the software does. It runs differently from the other popular online auctions, and although not very complicated, it is new.

The home user might be a novice user: a teenager, a college student, a senior citizen—almost anyone, including your mother. This user might be someone who has never used a computer before and found the program simply by stumbling across it during a Web search. Or the user just as easily might be an experienced computer user who enjoys an online auction and had your company's software specially recommended.

It's your job to make the program attractive and simple to use, because the first impression will be the most important to this user. In fact, if this user enjoys the E-Buy&Sell experience, he or she will recommend it to friends.

This user wants to know how to jump right into the product—in as little time as possible, and with as little effort as possible. This user wants to know how to work the product, not how the product works.

The Technical User

Same software, approach number two: Your assignment is to write documentation for a software developer in a client company. This client has bought the E-Buy&Sell auction software intending to put his or her own name on it, customize it, and run auctions independently. He or she needs to know how to do all these things.

The developer is not particularly interested in being on the bidding end of the software, and if he or she becomes interested, there's always the user guide. This user needs documentation to learn how the program works, and what the configuration files are. Most important, the user wants to know which parts of the source code and HTML files can and should be changed to make the program do what the client company wants it to do.

With this user, you can make the following assumptions: The user is technically savvy and therefore does not need a lot of explanatory detail, he or she is not going to use the program primarily as an end user, and he or she does not need a task-oriented approach. A developer uses *Application Program Interface* (*API*) documentation to understand how the program works, and how to make changes to the program.

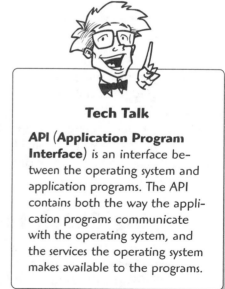

Tech Talk

API (Application Program Interface) is an interface between the operating system and application programs. The API contains both the way the application programs communicate with the operating system, and the services the operating system makes available to the programs.

The Big-Spending User ... Er, the Task-Oriented User

Suppose E-Buy&Sell's biggest customer is a brokerage house that will use the auction software to conduct stock transactions. This customer is very important and has asked for, and been promised, a customized guide that explains how to run this program to do stock transactions.

We're just teasing about "big spending." Although the importance of a customer might create a need for a special document, the most important thing to remember about this type of assignment is that to write good documentation, you need to learn about the business needs of this customer.

This can be the most difficult type of project you might encounter, but also one of the most satisfying if done right. Because the user needs are specific to a type of business, you must learn as much as possible about individual job functions, specific job tasks, and the client's overall business objectives.

Dodging Bullets

New tech writers often don't understand that some types of documents are more important than others. A writer will mistakenly expend the same amount of time and effort on a manual that goes to one customer, as he or she will on a manual that is widely distributed to all end users. Usually, a document that goes to *paying* customers is the one that deserves the most amount of time and energy, taking precedence over internal documentation or "one-offs" for special customers. And a document that is distributed to a large number of customers should receive more attention than a document that goes to a smaller subset of those customers.

Dancing Cheek-to-Cheek

After you have identified your users by asking others about them, the next step is to meet them. There are many ways to meet the people who will ultimately read and use what you write, and it can be fun—as well as enlightening—to do so.

Some companies are committed to usability and make it very easy for you to meet your user. In a supportive atmosphere such as this, you should plan to approach several people. Ask your marketing department which customers best represent typical users—these are the ones you should meet.

Coffee Break

When should you meet your users? Almost any time is a good time to make this contact. When you are just getting started on a document, it helps you know what type of material to include and how to write about it. In the middle of writing a document, user feedback can help you improve what you've already written and keep you on the right track. When your document is finished and has been released, meeting users helps you learn exactly what they liked and disliked, how they used your document, and how you can make it better for the next release. If you can establish and maintain contacts with users at all these times, so much the better—for you and for the user.

It's ideal to meet your users and interview them at their site, where you can see them actually using the product. By watching, you'll learn things that would never come up in a conversation alone. There are many ways you can have contact with your users, including …

➤ **The personal interview.** When you know you'll be meeting one or more of your users face to face, prepare by writing down your questions in advance. You can write a questionnaire and ask your users to complete it or simply write a list of questions that you will ask. It's a good idea to send the questions to your users before the meeting so they'll at least have the opportunity—if not the time— to think about them before you sit down together. Include questions the answers to which will help *you* write better documentation. Find out if the user reads your documentation at all, and if not, why not. If yes, find out what they like and why.

Pros Know

Be sure to thank your users for their feedback and insights! If you can, give them a gift for their help; maybe your company T-shirt or coffee mug.

➤ **Training classes.** Attending the training courses sponsored by your company helps you in two ways: It teaches you about the product you are writing about, and also it enables you to watch the users in action. There is a third value if the trainers use your documentation. You get to watch the students not only learn your company's product, but also read your documents! It can be quite an eye opener to watch a trainee fumble through a document you've written, searching for information.

➤ **Telephone interviewing.** If you can't visit your users, you often can arrange for a telephone interview. Prepare your questionnaire and go through it over the phone just as for a face-to-face interview. Same for videoconferencing, if that's an option for you. A conference call in which several users can be involved sparks dialogue and sometimes can reveal issues you hadn't thought of.

➤ **Mail-In questionnaire.** You might often see feedback questionnaires included in user documentation. As tools for collecting information, these don't work very well. The return rate on these is pretty low or even nonexistent. But that doesn't mean such questionnaires are without value. Some customers take them as evidence that you care about their needs and they like to see them, even if they don't send them back completed.

When There's No Face-to-Face Contact

In most companies you won't have the time, budget, or ability to meet the people who use what you write. Sometimes this is because the company is just developing its product and has no real customers yet (common in spin-off companies and start-up companies). Even in these situations, management has a good idea of a customer profile—after all, their marketing department has to target someone.

There are still some tricks you can use—or *workarounds* as we say in the trade. If you can't eliminate a problem, you simply "work around" it:

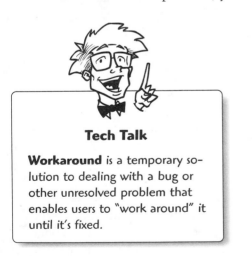

Tech Talk

Workaround is a temporary solution to dealing with a bug or other unresolved problem that enables users to "work around" it until it's fixed.

➤ **Become your own best customer**. If you have no way to meet your user, become the user yourself. As you use the product for the first time, make notes of the things that confuse you, and be aware that if they confuse you, they are likely to confuse someone else. Take lots of notes before you get too familiar with the product. Pretend you are doing a job with the product and note how intuitive it is—or isn't. Make sure you understand everything about how the product works and explain it to your end user as you wish it had been explained to you.

➤ **Find a proxy customer.** If you are writing for a specialized customer audience (one with needs and expertise well outside your own experience), find the equivalent person in your own company—a network administrator or programmer, for example, or a marketing person or installer. Ask him or her what the documentation should contain. Which leads us to the next challenge ....

What If the User Is Highly Technical?

What if you're expected to write a manual for a programmer? How can you write a manual for someone who knows more than you might ever know about the technology?

Your user can fall anywhere on the continuum between highly technical and hardly technical. Frighteningly, your user might fall well beyond your location on the continuum toward the highly technical end. And you might be asked to produce a highly technical document for this user, such as an API or a technical reference. What then?

Don't be frightened—tech writers have been doing this for years. Go to the person who knows even more than your potential user—for example, to the developer or developers who are creating the product. They not only will tell you what one of their peers would like to see in the documentation, but most likely also can supply you with that content.

Daunted by dealing with experts? Chapter 10, "Gathering Information," explains more about how to acquire technical information from those who know and how to assemble it into useful—and useable—documentation.

Remember, both your audience and the experts are busy people just like you. Keeping that single fact in mind will help you not only focus your writing for the audience but make the best use of the experts' time—so everybody wins.

The Least You Need to Know

➤ Know your audience. Find out who they are, what they do, and why they read what you write.

➤ Meet the user and ask him or her how you can make the documentation better. Then listen!

➤ If a user is highly technical, it doesn't mean that you can't write for him or her. Just go farther up the knowledge ladder.

➤ Remember that both your audience and the experts are simply people like you.

Gathering Information

Up until now you've been laying the groundwork: learning the product, defining your audience, determining the type of document that's needed, and developing a document plan. But you still feel as if you've got a tiger by the tail. Well, with a tiger that's the best end to start with. Likewise with writing a document from scratch (no pun intended). You know there's a lot more information out there; the kind of stuff you just can't find in books or marketing literature, and certainly not by hacking at it yourself.

Determining Content

"Content" is a shorthand term that refers to a broad aspect of a document; it simply means "all the information you want the reader to be able to learn by reading the document." It gives you a handle to use when you need to distinguish between content and another broad aspect of a document, such as format or structure.

Determining what content a document should have depends on two things: what your audience needs to know, and how they will use the document. For example, if your audience is still learning how to use computers, your content needs to include

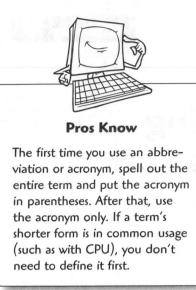

Pros Know

The first time you use an abbreviation or acronym, spell out the entire term and put the acronym in parentheses. After that, use the acronym only. If a term's shorter form is in common usage (such as with CPU), you don't need to define it first.

Pros Know

"I've worked in a consulting capacity for a lot of clients and with a lot of products, for Fortune 100 companies with big budgets and serious installations, and what I've found is that if a product has good documentation, my job is a lot easier. People can figure things out for themselves instead of paying me to do it for them. If two products are about the same, people will hate one and love the other because of the documentation."

—Jim Lemon, Consulting Systems Engineer, EMC², The Enterprise Storage Company, http://www.emc.com

lots of specific steps and "landmarks" (how the reader knows he or she is doing the right things in the right order). It might even need to lay groundwork such as explaining how to use a mouse or other pointing tool.

On the other hand, if your audience consists of network administrators or systems engineers, your content can focus more on information than on how to do every single step. After all, if your readers didn't have a certain level of knowledge and expertise, they wouldn't be doing those jobs.

The biggest thing that will determine content is whether your document takes a descriptive approach or a task-based approach to the product. If these terms sound familiar, give yourself a pat on the back—you had a brush with them in Chapter 8, "Learning Your Topic."

The difference between these two approaches—and it's a big one—is where the focus is. In the descriptive approach, the focus is on a "what," which is the product: What does it look like, what does each part or option do, what does the user do to manipulate an option or part? It's mostly up to the user to decide which feature or option to use for the result he or she has in mind. Often the result is not specific to the product. A digitized image, for example, can be created by many different products.

In the task-based approach, the focus is on the "why"—what a user wants to accomplish and how to go about it. The user depends on you to guide him or her in choosing among features and options to get the desired result. Often the tasks are closely or directly related to the product. An inventory management product, for example, might use a proprietary data format that makes updating the product database a unique process.

Description-Based: What the Product Does

With the description-based approach, every aspect of your topic is described clearly and accurately—if it's software, that means every screen, every entry field, every response display, every menu option, every taskbar, every dialog box. If it's reference material for commands, for example, that means every command and its options. If it's hardware, you might be identifying every component. You are saying to the reader, "Here is each part, and this is how each part acts."

Although at first glance the descriptive approach might seem to be the only way to be complete and methodical in writing about a product, that impression is deceptive. Why? Because people don't use computer products simply for the product's own sake. Most often people use computer products when they are trying to do something else: organize sales data, set up a network, manage inventory, exchange e-commerce documents, and so forth.

So why would anybody take the descriptive approach to writing a document? There are several good reasons. First, there's history. The first programs ever written were so rigid and limited that they could do only one or two things. If you wanted to do X, you used the X program; if you wanted to do Y, you used the Y program. In those cases, simple description was enough because the task was implied in the program. For some parts of the computer industry, this approach to documentation simply became a habit.

Coffee Break

Are you one of those writers who has strong feelings about consistency and how words should be used? If your organization doesn't already have a style guide, why not go ahead and offer to write one? Doing so has the major benefit of letting you have a big say in decisions that may otherwise drive you crazy, and you're probably the one who cares most about it anyway. So do it. You know you want to.

Second, there's human ingenuity. In some kinds of development software, such as for animation and image manipulation, the number of possible things a user can do with a product is almost infinite. When a user's original vision is what drives the task at hand, describing the tool itself accurately and completely is what the user needs most. The user then can match the tool's function to what he or she is trying to achieve at that moment.

Third, there's cultural literacy. In every culture (or subculture) there are some things that, once learned, don't change, such as how to dial a phone or use a calculator. Once you learn how to use one phone, you're more or less equipped to use any of them. A simple descriptive list of phone numbers becomes meaningful and adequate because you already know exactly what to do with them. The same situation exists on a smaller scale in specialized fields from computer graphics to microcircuit design; sometimes a clear and complete product description is all the user needs.

Consider using the description-based approach if the task is clearly implied in the product, if the possible tasks are so many and varied you couldn't possibly describe them all, or if it's safe to assume your user is already familiar with that type of product and only needs the details about yours. You might describe a few tasks, but your content will be mainly about the product, down to the tiniest nuance and detail. You'll know the product inside and out when you're finished.

Coffee Break

In the 1960s, doctors discovered while working with patients with severe epileptic seizures that each hemisphere of the brain processes information differently. The left hemisphere is dominant in verbal, analytic, abstract, and logical activities. The right hemisphere is dominant in nonverbal, analogic, intuitive, and spatial activities.

Which side do you think is dominant in most tech writers? A lot of people get into technical writing because they are logical, analytical people who work well with other logical, analytical people—clearly left-hemisphere dominance. But a writer also has to dig into the right side of his or her brain when it's necessary to approach an idea visually, to design a document for maximum effectiveness, or even to decide which of several approaches will appeal to the user.

There's one last reason for using a description-based approach—that is if you are limited to producing only one manual for a product. If budget or other restraints mean that your hardware or software will come with only one document, then a full explanation of every aspect of the product may be the way to go. Anything less, and your user will probably be searching for just that missing piece of information.

Task-Based: What the User Does

With the task-based approach, the focus is on specific results the user wants to achieve and how he or she should go about getting them. This requires a greater understanding of the user's typical job duties, and possibly how those job duties support certain business objectives.

For example, your company might sell a product that helps Web-based businesses track online sales figures for a wide range of products. Perhaps a key feature is the ability to push that sales data to the CFO's cell phone every 15 minutes. The user's tasks then become "setting up continuous data reporting" and "pushing data to wireless devices." There are no ifs, ands, or buts about what the task is, whether it needs to happen, or exactly how it should be done.

With a task-based approach, your focus ultimately is on the user's objectives and how the user pursues them. Your content, then, must encompass not only the product itself, but a larger understanding of how the product is used.

A pool of quicksand that technical writers sometimes unwittingly step into, however, is trying to write instructions for a task when what is really needed is simply a description of a process.

Two things determine whether you're dealing with a true task: whether there's a clear, known, predictable trigger that starts it, and whether there's a clear, known point where it stops. Ask yourself these questions: What starts this process? Is it clear who or what is acting at every point? How do I know when it's finished? If any of these points is not clear, a description of what actually is known is the best you can hope to do. An interesting side benefit of using a task-based approach to documenting a product is that it can reveal design flaws. How? A well-designed product should have features and options that are known to be of value to its intended users. Every function, feature, and option needs to be active in some task the user performs.

You also might discover things the product can do, but be unable to discover why it does them. Often design requirements change during development and obsolete features from the original design aren't always removed; problems initially tackled one way but ultimately solved another might leave traces behind. And it's quirky but true that sometimes developers make products do things just because they can.

Bear in mind that although it's not up to you to decide if a product design is right, it is your job to examine the obvious and to ask appropriate questions.

Gathering Information

Collecting the information you need for your document is a lot like a treasure hunt: You start out with clear objectives in mind, and you do your best to meet them, but you also might discover some surprises along the way. Plan to write down everything you hear, see, or even think that relates to the topic of your document.

Keep your information in a notebook or something that enables you to keep it all together. As you collect information, try to categorize it—installation, configuration, user, troubleshooting, or whatever category it seems to fall under. Carry this notebook with you throughout the life of the document, and when that document is finished, start another notebook for the next document.

When you begin gathering information, work from the broad to the narrow, from the general to the specific. Start by defining working labels for big conceptual "chunks," then explore each respective chunk. And revisit those marketing materials you got earlier, when you were learning about the product; you might be surprised how much more they tell you now that you know more.

Task-Based Documents

If you're doing a task-based document, you'll start by coming up with a working list of tasks to be documented. Make this list as specific and definitive as you can at this point, bearing in mind that it's likely to change over time. Roughly draft the steps you already know of that make up each task, and clearly indicate the places you aren't sure about.

Dodging Bullets

When you sit down to make notes, be prepared: Make sure you have plenty of paper, extra pens or plenty of lead in your mechanical pencil, and high-lighters or colored pens to mark especially important points and paragraphs. Being ready is partic-ularly important if you're work-ing with a colleague whose time is at a premium. Running out of supplies not only wastes time; it also makes you look unprofes-sional.

Description-Based Documents

If your document will be description based, begin by requesting a copy of all the product's screens and dia-log boxes, if it's software. It's a good idea to print out hard copies of the information, as you'll want to make notes on them and mark them up in other ways to draw attention to specific parts.

In either case, it will help you to have a copy of the product's design specifications. This is a document produced very early (at least it should be) in the devel-opment process that spells out what the product is in-tended to do as well as its limitations. Design specs also often include the names of key players in the de-sign process, which can be a big help in getting ques-tions answered as your document develops.

If there's no formal design specification (which some-times happens in hectic startups), ask about obtaining copies of informal notes or hand-drawn diagrams. Sometimes you can get them, and they can be helpful for filling in background for your own knowledge, if nothing else.

Of course, your best source of information about the product, after the product itself, is the people who de-signed and developed it. These are your Subject Matter Experts (SMEs). And you'll get that information from them the old-fashioned way: the interview.

Interviewing SMEs

Relax, nobody expects you to be Barbara Walters or David Letterman. The interview is simply two (or more) people getting together to talk about something they have in common: the product.

If you haven't already been introduced to the people you'll speak with, it's a good idea to introduce yourself or, better yet, ask their managers to introduce you. That way you won't be a complete stranger when it's time to get down to business. It's also important for them to understand that spending time answering your questions is something they're expected to do, rather than being an intrusion on their already busy schedules.

"Interview" is too formal a term to use when you're setting up times to get together, but do be specific about when you want to meet, and for how long. It can be as simple as saying, "I need to sit down with you and talk about this new product for an hour or so tomorrow. What's a good time for you?"

Making a Good Impression

If you're in the position of speaking with a SME for the first time when you call to set up a time to meet, be sure to introduce yourself and explain how you fit into the project before you ask to get together. Name-dropping might be frowned upon in social circles, but in these circumstances it can help establish your bona fides with someone who's never met you and might not yet know he or she will be working with you.

The question of where to meet can be easily resolved. Normally it will have to be at the developer's or engineer's desk, where he or she can show you the product and access notes and other materials.

On the other hand, sometimes it's impossible to exchange two sentences in a row at a busy programmer's desk. You'll be interrupted every third word by the phone, a pager, the "ding" of arriving e-mail, or colleagues dropping by. In those cases, it's best to schedule a meeting room where you can have his or her attention all to yourself. (Be sure to close the door. It's not unfriendly; it simply sends a clear signal that what's going on in the meeting room is important and should be respected.)

Pros Know

A small tape recorder is an invaluable tool when interviewing a SME. It later enables you to remember what was said, and you can play it over and over without annoying the speaker with repeated questions! Best, the tape recorder frees you from taking notes, which can require so much work and concentration that you might miss the gist of the conversation.

Keeping It Casual and Friendly

When you arrive for the interview, set a casual, friendly, but professional tone at the outset. Everybody will be more relaxed (including you). Lots of people have never been "interviewed" before and they might feel awkward about it at first. Just stay confident and relaxed and they'll soon loosen up.

Be prepared for your time together by having a set of questions ready. If you can, send your questions to the interviewee ahead of time so he or she will have a chance

Dodging Bullets

You'll find you often need to be persistent—especially when you're chasing that elusive SME. Sometimes, no matter how many e-mail messages you send or how many drafts you put on someone's chair, he or she simply won't respond to your questions. If that happens, try to get the information you need from someone else. If there is no one else, as a last resort, ask your manager to intervene.

to think about them. This can save valuable time for both of you. You might even get answers to some of them before you meet.

Keep your questions focused and on-topic; but this isn't the Spanish Inquisition. A little light humor at an appropriate moment can go a long way toward building good working rapport.

Ask Intelligent Questions

If you have learned about your subject matter, as discussed in Chapter 8, make sure your questions reflect this. Don't ever walk into a developer's office and ask a sweeping question like, "How does this product work?" Ask questions to which you can expect concise answers, which show that you have done your homework.

Know When to Listen

Perhaps the single most important part of interviewing someone is knowing when to be silent. Give your interviewee a full chance to answer each question. Don't expect quick, immediate answers to your questions, and try not to anticipate what they'll be; thoughtful responses take time to formulate, so give the other person plenty of time. It's tempting to interrupt when you think you know the reply, but courtesy is more important than speed. Nothing can derail a productive interview faster than interrupting the interviewee or trying to anticipate what he or she is going to say. If you think you already know the answer, it's still best to listen: if you're right, it's confirmed; if you're wrong, nobody needs to know but you.

Keep an eye on the clock and acknowledge when your time is nearly up. If you need another session, that's a good time to schedule it. The fact that you've respected the meeting's stated limits often makes it easier to schedule follow-up sessions.

Do a Quick Review

Or, if you've gotten what you needed, now is a good time to review your notes and questions and fill in any gaps or clear up any points you weren't sure about. You also might want to ask if there's anything the other person can think of that you should know but that you didn't ask about. The response to this question can surprise you—mostly it will be, "No, I don't think so," but you could hear, "Oh yes, I need to tell you about …." Asking this question can present you with an essential piece of knowledge you might never have discovered otherwise.

Keeping Up

If you are watching a developer or expert user take you through a product step by step, you are at a disadvantage: You must slow down your demonstrator, and note not only what's happening on the screen but where the mouse is pointing or what the hands on the keyboard are doing. This makes reducing speed very important.

You'll find yourself repeatedly asking the demonstrator to go back and show you something again or to tell you how something was done. Don't get tense or apologetic about this; you're doing your job, and the demonstrator's job at that moment is to help you.

But do make an effort to keep things moving. And don't wait until the end to say thank you. Let your colleague know you appreciate the extra effort it takes to go slowly. Most developers are very busy, and however gracious they are, they don't like to interrupt their work to talk to a tech writer. Some developers resent the process, feeling that the tech writers expect too much input from the developers. Your obvious consideration for their time and workload—and the fact that you've taken some time to learn about the topic—can ease resentment or prevent it in the first place.

Pros Know

It can be annoying for SMEs to make the frequent pauses you need to write good notes. After all, they're busy people, used to working at top speed. But they're also human, and sometimes a solution is for him or her to eat lunch—or some other missed meal—while the two of you are working together. (Eat before you meet so you can concentrate on your notes.) This can make it possible to meet with an expert who otherwise might not be available.

Solve this potential problem by keeping visits short. Instead of scheduling (or expecting) a long session with a busy programmer, ask one or two questions at a time. Submit small sections of documentation to see if you're on track; and, if possible, wait while he or she reads it. Also send your questions in e-mail; sometimes a busy developer answers his e-mail at home or after the workday ends.

After each session with a developer, it is a very good idea to go back to your desk and immediately add what you've learned to your outline or draft. If you don't do it right away, other things distract you, and one day you pick up your notes and can't understand a word. It's funny how that happens, but it happens all the time if you don't organize your material while the information is fresh in your mind.

Patience and Persistence: Eyes on the Prize

No discussion of information gathering would be complete without mentioning the value—the absolute necessity—of both patience and persistence. You're sure to need them both, and often.

Coffee Break

How long is "now"? Research indicates that most people experience "now" as being about three seconds long. Keep this in mind as you chase down that elusive fact or grapple with a resistant graphic—you only have to do it for three seconds at a time!

Why? It has to do with the nature of the work. Technical writing is a continuous process of carefully gathering, sifting, organizing, and assessing. Each of those steps takes patience on its own, and you have to do them one after another—often more than once. There's no word other than patience to name what will get you through that.

Persistence comes into play in several ways when writing technical documentation. Not only do you need it to keep firm hold on the vision you're shooting for—what the finished document will be like—but also in pursuing the pieces needed to make that vision real, step by step and day by day. Technical writing is not for wimps.

Don't get us wrong—you're not doing constant battle with the forces of chaos, evasion, and delay. Most people honestly want to help you deliver a high-quality document that's accurate and complete. But the pressures of the computer industry are many and unrelenting, and so are the demands on everyone's time and attention (including yours). The focus you bring—in the form of patience and persistence—is one of your most valuable assets.

The Least You Need to Know

➤ Decide early on between a description-based (focuses on the product) or task-based (focuses on the user) approach.

➤ Treat knowledgeable people with friendly respect—they are your greatest information resources.

➤ Use the organizing process to focus document content more clearly.

➤ Don't take it personally if information is hard to come by; just be patient, persistent, and polite in your pursuit of it.

Part 3
Racing Toward the Finish

This part speeds you through creating your document, from the first rough draft through handing it off polished and complete.

This phase of document development truly is a mixed bag. One day you might be conducting a table-top review, working closely with colleagues to hammer out just the right content and organization, and the next day might find you working alone for hours, creating your document's index.

But this is one of the appeals of tech writing: the wide variety in activities from day to day. Learn to love them all, or at least like them enough to do them well.

Chapter 11

The First Draft

In This Chapter

➤ Creating an outline

➤ What a first draft needs—and doesn't need—to be

➤ Writing your draft

➤ Some tips for cracking through writer's block

You're about to earn your wings—water wings, that is. You've got a handle on the expectations for your job, you know what a good document looks like, and you know how to write clearly and appropriately for a technical audience.

If you're part of a documentation team, by this time you've probably tested the waters by updating a manual or two or editing documents written by someone else such as a programmer or system architect. If you're the lone tech writer in a department or at a small company, however, you might be facing this leap off the high board before the end of your first week on the job. Either way, the assignment is the same: Write a first draft from scratch for a new document.

At this point, it might feel as if you've jumped off into the deep end. But don't worry—the water's fine. This chapter tells what is (and isn't) realistic to expect a first draft to be and do, how to take that outline to the next level, and about how long that first draft should take you. Take a deep breath and get ready to swim.

Jump Right In, the Water's Fine

We assume you've already done the basics such as learning the product, defining document scope, and gathering all your information. You're ready to write—even though you might not feel like it. Here are some basic truths about first drafts that we hope will help the drafting process go swimmingly.

Coffee Break

Beginning writers often have expectations that are, well, just too ambitious for a first draft. In an ideal world, you'd write a single draft of a document, get it reviewed, make a few minor changes, and be finished. It's true sometimes things happen that way, and it's the ideal to aim for, but in real life it's the exception rather than the rule.

The process of writing a first draft is, frequently, a continuation of a discovery process instead of simply organizing and cataloging known facts. Typically there are many changes right up to the end of the schedule. Be prepared for this so it won't take you by surprise.

It doesn't mean you're disorganized if you don't write the whole document straight through in sequential order from the introduction to the last appendix. If you're drafting this document while product development is still underway, you might have to start with the part of the product that is most complete or most stable. You might end up writing the introduction last instead of first. As long as all the parts fit together in the end and the document has a sense of unity, it doesn't matter in what order you wrote each piece. This is where the outline helps enormously. If you've prepared an outline, even sketchily, you can easily figure out what goes where and what's missing.

In the Beginning: Creating a Document Outline

Creating an outline for your document is a lot like starting on that jigsaw puzzle; different people start in different ways. Some gather all the edge pieces first and build a frame to fill in; others sort the pieces by identifiable patterns and colors and group them in areas where they seem to belong. The content you've gathered is a lot like a box of jigsaw puzzle pieces dumped out on a table: The whole picture is there somewhere—you just have to organize the pieces to see it.

Your outline will be much different for online documentation than it is for sequential documentation. A printed manual usually has a beginning and an ending, with some kind of progression in the middle. A set of online documentation has a tree of information that must be tracked carefully to make sure that the user can navigate from one spot to another easily.

Online documentation requires more detailed outlines than the sequential type does. If you are drafting hypertext documentation for the Web, refer to Chapter 22, "WWWriting for the Web," for more information.

Outline Format

Classic outlines use capital letters to indicate the biggest headings, then numbers, and then lowercase letters. You can continue to build the outline by alternating numbers and letters again. Each item has a different indent. The farther in the item is indented, the more specific and detailed the information, as shown in this sample:

A. Introduction

    1. About E-Buy&Sell

    2. About This Manual

        a. Typographical Conventions

        b. Contents of This Manual

    3. System Requirements

        a. Hardware Requirements

        b. Software Requirements

        c. Color

        d. Optional Add-Ons

B. Getting Started

    1. Introduction

    2. Keyboard Shortcuts

    3. E-Buy&Sell Tutorial

        a. Logging In

        b. Creating a Window

        c. Making a Bid

C. The E-Buy&Sell Interface

    1. Introduction

And so on, as deeply as you can or care to go. The classic outline format is fine, and so is any other method you use as long as it works. You might work best with a very sketchy outline of just some headings, or you might be one of those super outliners who continues to fill in more and more information at larger indents and at more detail. If you can continue to progress in this manner, you actually can write your entire document this way, and some tech writers have!

Start your outline by writing all the chapter headings you expect you'll need. Use other documents as reference by finding a document close to the one you're writing, and copy all the headings that seem appropriate. You can always delete them later if they aren't right. For example, as you refer to your outline later, you might decide that you don't want a chapter called "Introduction" and then a major heading called "Introduction" in each chapter. Laying everything out in an outline helps you to clearly see both relationships and repetitions of information.

Shuffling the Virtual Index Cards

People used to create outlines by writing all their topics on index cards and shuffling them around in huge piles, laying them in places where they seemed to fit and moving them if they weren't right. Luckily, we now have the technology to cut and paste. Use your computer's word processor—it takes up much less floor space than a big spread of index cards.

The method is the same whether you use index cards or a word processor, though. Whatever way you start your puzzle—er, outline—the goal is the same: to place each piece where it fits best in the whole.

Refer to the notes you made during your information gathering expeditions and start plugging the information into the appropriate places. If you're at a loss for how to begin, here's one good way:

1. Divide your content up into groups or "chunks" that are collections of related pieces and processes. Don't worry about order at this point. Focus on relatedness.

2. Start at both ends. What content chunk is introductory or foundational? Put it first. What chunks are supplementary or not at the content core? Put those last.

3. What's the first thing a user needs to do or know? If there's no clear sequence you can see, take a look at frequency. Is there one thing readers will use or do more often than others? How about a shared action or aspect before options diverge? Put that next after the introductory chunk or chunks.

4. What's the next thing in the sequence, or the next most frequent action? Repeat this step until you've accounted for all your chunks.

If you find yourself struggling with a chunk that just doesn't seem to fit, take a second look at that chunk. Perhaps the unifying idea isn't quite on target and needs to be revisited. Maybe that content needs to be more than one chunk, or shouldn't be a chunk at all. The parts that initially seemed to belong together might really work better distributed among the other chunks.

Another strong possibility is that you've identified a chunk of content that is worth keeping but that simply doesn't belong in the document you're currently working on. Set it aside for now; you'll probably want it for a different document later.

That's another value in the organizing process—in addition to shaping what you'll include, it helps eliminate parts that don't fit. Your current document comes more clearly into focus, and reading what you've written can show whether there are gaps in your knowledge that might be tripping you up.

Drafting Your Draft

The first paragraphs or pages you write about the product might end up being thrown away—and that's okay. This is a draft, not a finished document. Of course, you want the draft to be as close to its final form as possible (typical schedules being what they are), but remember that writing is about the process as well as about the results.

Sometimes you have to start getting words up on the screen or down on paper before the ideas become completely clear. The important thing is to write—you can always rewrite, revise, or reject what has been written, but you have to have something to work with in the first place. You can't correct what doesn't exist.

Remember that notebook we recommended you keep in Chapter 10, "Gathering Information"? Now's the time to make sure you use it. Go through the notebook and methodically deal with every piece of information in there. Either add it to your draft, or make a decision to not use it. Cross out each piece of information as you add it to the draft.

Expect that you will have a lot of questions you can't answer. It's even a good idea to include them right in the text, as notes for reviewers or yourself, like this:

> Remove any files you don't need, such as core files. [**What else should they remove here?**]

That way everyone can see exactly what's missing or what's not clear, and the context the information needs to fit. Embedding questions in the text

Pros Know

Embed questions right into the text of the first draft and set them off with special characters or a different font—you might want to **>>frame them like this**<<, or [PUT THEM IN BRACKETS WITH ALL CAPS LIKE THIS], which makes them easy to find with your word processor's or DTP's "find" feature. Or, if your documents will be reviewed electronically instead of on paper, you might assign a color to use only for questions. Embedding questions in the text gives reviewers a context for the missing information.

also emphasizes that it's a first draft, not a final version, and provides an important visual cue for where reviewers need to focus their attention.

Aside from embedded questions you should provide other unmistakable cues that the document is a draft—let the word "draft" appear prominently in the header or footer, or as a large shaded watermark image behind the text on every page. This protects your company as well as alerting reviewers—when draft documents that contain known inaccuracies merely as placeholders get unofficially copied and distributed, the results can be ugly and expensive. With a clear draft indicator visible on every page, nobody can reasonably mistake a draft for a finished release.

It's not impossible for draft documents to become targets of industrial espionage, either. Such cloak-and-dagger concerns might seem farfetched at first, but if you consider the value of the specific content collected in a technical document, you'll soon see why a competitor might be eager to have it. Beyond basic vigilance, it helps to clearly identify proprietary documents—at best, it will encourage others to handle them appropriately; at worst, it will give the legal department a leg to stand on in asserting that the information was not intended for distribution outside the company. Other indicators of draft status lend weight to this argument as well.

The Hipbone's Connected to the Leg Bone: Fleshing Out the Outline

To start writing, choose the chapter or section of your outline that feels most complete or about which you already know the most. If there's no clear favorite, just start at the obvious place: the beginning.

If the chapter or section is going to be a long one, you might need to begin by creating a separate outline for it before you start writing. This can save valuable time by keeping your efforts focused from the beginning. Chapters often have similar structural components, such as an overview that starts things off followed by topic specifics or procedures. Use these to anchor your section or chapter outline.

But if the existing outline seems like enough, just start writing. Not sure how to start? Making a declarative statement about the topic at hand is a good way to get momentum going—"The XYZ Product is …" or "There are X main components to the XYZ Product." Another way is to state the main idea of the paragraph, then follow that statement up with details in subsequent sentences. This works with any paragraph, not just the first ones.

A good analogy is adding clay to the underlying support structure of a piece of sculpture—stick a blob on here and hunk on there until the piece begins to take shape. When the "sculpture" is finished, nobody will know (or care) which blob you added first.

Some writers prefer a more methodical approach—like hand-dipping a candle rather than adding clay to a sculpture. For each heading on an outline, add any subheads that seem necessary. Then, for each heading (or subheading—whatever is the lowest level), make a bulleted list of phrases that capture the essential points to cover under that heading. When every heading has a bulleted list, start at the top and write the text to account for each bulleted point.

As you can see, with this method each heading of the outline starts developing depth at the same rate. The "candle dipping" approach is helpful when there's a high volume of content to be managed—it enables you to make structural adjustments before you've written a lot of text.

The "candle dipping" approach also is useful if you know there's a good chance someone else will have to write from your outline or that you may have to set it aside and come back to it later. The document doesn't end up with uneven content because things that earlier were in your head but not on paper got forgotten later.

Another way to start fleshing out an outline is to take parts from documents that already exist for this product and drop them into yours. Don't be shocked—it's not plagiarism to incorporate existing text from other documents your company has already published. On the contrary; this is called "leveraging existing materials" and can save you a lot of time and effort when such documents are available. Graphics, tables, and even whole blocks of text from marketing materials and specification sheets can fit perfectly (with perhaps a little massaging) in lengthier documents as well.

And, you have the added benefit of knowing the content has already been approved. Check the dates on the materials, though, to be sure the content is still current—perpetuating obsolete information isn't helpful to anybody, least of all the user. But even if it's out of date, published information gives you a solid place to start finding out what's right.

Dodging Bullets

A word to the wise: You might find that perfect paragraph, turn of phrase, or to-die-for graphic in a competitor's published materials or Web site. Although it might be tempting to lift text verbatim or copy a graphic image, resist, resist, resist. Not only can that make you and your company look bad, it opens your company (and possibly you personally) to charges of copyright infringement.

How Fast Is "On Time"?

In addition to juggling flaming sticks (see Chapter 3, "The 'Write' Stuff"), you also have to do a balancing act—balancing the time you have available to write a draft and the document deadline you're working to meet.

Keep in mind the function of a first draft—not to dot all the i's or cross all the t's but to put down the content as you *think* it is, to get confirmation on the document's basic structure, and to elicit answers to lingering questions. Writing a first draft is like applying a first coat of paint to a wall—put on enough, and then stop.

Chapter 17, "You Want It *When?*" gives some guidelines about the amount of time you might take to prepare a draft and the amount of completion you should expect from it. Think about aiming at criteria while that first-draft clock is ticking: knowing it won't be complete and it won't be 100 percent accurate, work to make your first draft as complete and accurate as possible in the time you have. It might feel like working without a net, but have faith—the draft will turn out fine.

Battling Writer's Block

When deadlines are looming and the words just don't come, try some of these tips for getting your writing started:

Block out some time. And we do mean block. Turn off the phone, close the door (if you have one), and give yourself time to work. We're talking time in increments of hours, not minutes, here, as many as you can set aside. Can't work at your desk without being constantly interrupted? Find a conference room no one is conferring in and use that (remember to close the door). Or work from home if your management agrees and you have the equipment. One mistake writers sometimes make is to think they will be able to sit down and knock out some content during a free half-hour break. Maybe a few people really work this way, but in our experience it takes more than a half-hour to gather serious writing momentum.

Maybe it honestly takes you 45 minutes of dawdling, drinking coffee, and reading e-mail to warm up before you can hammer out three solid hours' worth of work. You have to respect that—but don't let a 45-minute warm-up turn into the whole day.

Just do it. "But it's not that simple!" you exclaim. We think it is. It does take discipline to start and continue working, especially at the task of writing, which is full of stops and starts. And yes, staring at a blank computer screen or sheet of paper is daunting. But waiting for inspiration is not an option. Just put your hands on the keyboard and go. There's no substitute for simply sitting down, saying "Now," and getting to it.

Start anywhere, with anything. Sequential thinking is a great ally when you're working through a process or organizing Web pages, but there are times when it's not your friend. If you want to begin at the beginning, but find yourself chewing on a pen and unable to figure out where to start, start anywhere. Pick up a marketing brochure or a developer's specification and retype it. The act of putting fingers to keyboard often is all it takes.

Rewrite and edit as you type it, and go back later to edit it again. List five things about the product that the brochure or spec doesn't mention. Eventually you'll discover you're writing your own ideas in your own words. You've made a start.

Dare to be bad. Maybe you're afraid to write something until you can feel confident it will be good. Forget that! Start with facts you know are true and write *anything* with them. Never mind how bad the writing seems to you at the moment—choppy sentences, no transitions, a handful of unrelated statements jammed together to make a paragraph. (Do, however, stick with accurate content—these "bad" early paragraphs have a funny way of showing up in the final document.)

Each piece you write doesn't have to fit anyplace right now. Don't worry about that—you can find a place for it later. Write 8 or 10 of these chunks of text, then look for themes. You probably can start to string them together as the core of a document.

When you read the early paragraphs later, you might even discover they aren't so bad after all; in fact, they're probably pretty good. But even if they're not, they still did their job of getting the writing going. Edit each piece or delete it, pat yourself on the back for daring to be bad, and move on. Now you're rolling.

Make an outline. We've told you that already. But if you've got this bad a case of writer's block, it's possible you didn't make the outline. Once you have an outline, you can add information to each section and then string all the parts together later.

For the terminally blocked, even this tactic might be too much. If your fingers freeze into immobility when you even think the word "outline," creating empty chapter files with titles and some headings in them is still better than doing nothing.

No matter what it takes to get you going, you're really only a few hours away from that first draft. So start writing!

The Least You Need to Know

➤ Create an outline. It's a great way to capture and organize what you think should go into the document.

➤ Be reasonable—have realistic expectations about how complete and accurate a first draft can actually be.

➤ Rely on your outline to keep your writing on target.

➤ Flesh out all parts of an outline evenly if others might work from it or if there's a chance it might be temporarily set aside.

➤ Balance time and results—there's only so much you can do in the time available, but do all you can.

➤ If writer's block sets in, the best thing to do is ... write anyway (even if what you write is terrible)!

Chapter 12

Everybody's an Editor

In This Chapter

➤ The importance of feedback

➤ Working with reviewers: who, how, and when

➤ Coordinating feedback from a host of reviewers

➤ How many review cycles are enough

You've finished the first draft of a new document or made what you believe are all the required changes to update an existing document, and you've sat back with a sigh. But don't get too comfortable—you're not finished yet.

The next phase is about to begin: the feedback process. This is where you distribute the draft (a revised document also is a draft—even a second, third, or fourth draft— until it receives final approval) to a chosen group of knowledgeable people. It's their collective job to confirm that what you've written is true and that nothing has been left out. These are your reviewers.

In this chapter we'll guide you through this sometimes challenging aspect of technical writing. By the time you reach the end of the feedback process you indeed might feel as if "everybody's an editor!"—but you'll also be grateful for their help in making your document as accurate and complete as possible.

Getting Feedback

If you've done your homework and double-checked your facts, you might feel that asking for feedback is a mere formality. After all, you're sure what you've written is right, and you included all the information you believed was needed. But there are still some compelling reasons not to skip the review process—and some specific things you can do to make it go smoothly.

Confirming Information

At the top of the list is confirmation that what you've written indeed is accurate and complete—and current. This last one is what can trip you up. It's not at all uncommon for changes to be made to a product that you don't know about—maybe they've been added at the last minute, or someone dropped the ball on keeping you informed. For whatever reason, what was true and complete when you learned it might not be so today. The review process is the only surefire (well, almost surefire) way to have the highest possible confidence in your document.

Tech Talk

Stakeholders are the people who have a stake in seeing that your document is accurate and complete. It's another way of expressing a sense of shared responsibility. People who will be affected by a document's failure or success care more about it than people with no stake; so they put forth greater effort. Creating stakeholders not only gives your document a better chance of being its best, it spreads the responsibility for it over a larger group.

Showing What's Still Needed

Another reason for having your document reviewed is that it demonstrates exactly where holes still exist in your information—not only can reviewers see exactly what's missing, they can see why it's needed and where it has to fit. Developers don't always understand why you want to know something, which means their responses to your requests can be less than thorough. When they see a gap in a draft document, they suddenly "get it" about that question you've been pestering them to answer.

An especially compelling reason for document reviews that seldom receives mention is that of building *stakeholders* for the document. This means that a larger number of people share responsibility for a document's completeness and accuracy. Signatures of approval from developers and managers mean they are putting their best judgment behind the document along with yours.

Not only does this ensure the information's integrity, it means you won't be left holding the bag if a problem or oversight emerges later. This aspect is especially important when you are writing about technologies or applications outside your field of expertise where an innocent misinterpretation could turn fact into fiction.

With all these important reasons to have documents reviewed, you can see how important it is to also lay the right groundwork—with yourself as well as with the reviewers. For yourself, let go of any emotional investment you might have in what you've written. This is not the product of your soul's labor, birthed from your heart in a flood of blood and tears. It's a technical document—a collection of facts, descriptions, and procedures. And it might sound silly to say it, but this document is not *you*. Its flaws and shortcomings (and yes, it has those) do not reflect on your value as a person or your capability as a writer.

A surprising number of tech writers haven't learned this fundamental "thick skin" lesson—which is one of the primary differences between writing for yourself and writing for hire. You simply can't take it personally when reviewers start cutting apart a document you've written—and cut it apart they will, which is exactly what you're asking them to do. You want them to push it, pull on it, point out its weaknesses. That's the only way you'll know how solid it is and how to make it better.

And this brings us to the topic of educating the reviewers before that first review. There are three things reviewers probably don't already know, but need to. It's up to you to tell them:

➤ **The documentation development process.** Most SMEs have no understanding of what happens before, during, or after the review process they participate in. It helps them to help you do your job if they see how their pieces fit into the puzzle.

➤ **The difference between a "review" and a "critique."** In a technical writing context, a review is an intellectual exercise that focuses on document content and how it is organized. Does it make sense to address this topic here and that topic there? Is this all that needs to be said about a feature, option, or process? A critique, on the other hand, is a

Dodging Bullets

Reviewers have met some oversensitive tech writers and often are afraid to say anything that might be taken as critical or negative about "your" document. Give them some help: Lean forward, look interested but not apprehensive, and listen calmly as they tell you what's "wrong" with the document. Then surprise them by smiling, nodding, and saying, "Okay, good, thank you!" and make a note of their point. Then ask "What else?"

Pros Know

If you're not sure who should be a reviewer, make sure you include the manager of the development group responsible for the product, the key developer or developers working on the product, and your boss. Then try to choose at least one person from each of these areas: marketing, technical support or *quality assurance (QA)*, customer support, and documentation.

Tech Talk

QA (Quality Assurance) is a set of procedures that includes the entire development process: monitoring and improving the process, making sure that agreed-upon standards are followed, and ensuring that problems are found and resolved. A company may or may not have a separate QA department responsible for carrying out such procedures.

Dodging Bullets

When choosing your reviewers, be aware of people who might need a chance to see and informally approve your document but who aren't necessarily experts in its content. Depending on your company's culture, these might be members of upper management or the heads of other departments or other divisions. It's important to not accidentally step on someone's toes—and it's easy to avoid it. Ask your supervisor if there's anyone else who might need to see the draft you're preparing.

person's subjective judgment about the "goodness" or "badness" of something—what the critic liked and didn't like. This information can be useful if you can get at the underlying reasons for a preference, but more often this kind of feedback is simply off target.

➤ **It's okay for them to tell you what's "wrong" with the document.** Many times, reviewers don't want to "hurt your feelings" by "criticizing" what you've written. They need to understand that your feelings won't be hurt and that you want to know about errors so you can fix them.

Choosing Reviewers

Choose reviewers whose input will be meaningful—which means the information will not be news to them. On the front lines to review a document are the people who provided the information that's in it. Ideally these should be the people who know the most about the product anyway, and this is the perfect opportunity for them to check and confirm that nothing got lost—or distorted—in translation.

In building your list of reviewers, think also of whether there are other departments that need to be aware of the document's content or that might have unique insights to offer, such as sales or marketing. People in daily contact with customers or clients can provide input nobody else has.

You don't need to worry much about covering the bases with management review and sign-off at this point unless you are very sure about the content, as in the case of a superficial update of an existing document. See "Once More, with Feeling" section later in this chapter, for more about additional and final reviews and sign-offs.

Be sure all your reviewers know ahead of time that the review draft is coming, and that (if they have a choice) they agree to participate. Doing a thorough review of a document that might be a hundred—or hundreds—of pages long takes a significant investment of time. People do not appreciate being surprised with this kind of weighty extra task.

Preparing the Draft for Review

Making the review process easy for the reviewers goes a long way toward getting the feedback you want and need. Make sure your draft includes the following attributes:

➤ The word "DRAFT" on every page in an obvious location

➤ The date on every page

➤ A draft number or revision number (for a first draft, 1.0)

➤ Correct information in the header(s) and footer(s) (document name, chapter or section name, proprietary statement, copyright statement, and so forth)

➤ Screen shots or graphics that are either as correct as you can make them, or that are clearly identified as placeholders (labeling them "PLACEHOLDER" might seem *too* obvious, but it isn't)

➤ Embedded questions marked in an eye-catching way that unquestionably differentiates them from surrounding text

It's important to address all of these whether the draft will be reviewed in print on electronically. Why? Because once the document leaves your hands, you can't control what happens to it—and review drafts have been mistakenly (or perhaps not mistakenly) copied and distributed to a wider audience before.

The draft needs to have unmistakable markers to indicate its draft status and when it was created. It's surprising how review copies of things seem to take on a life of their own. They'll resurface unexpectedly weeks or months down the road—a sales rep will give a draft copy to a customer, or a member of your installation team will take a draft copy to a customer site and leave it there. More than once, a client has received an unmarked draft and later become angry when it turns out to be wrong, or when a newer version is released officially.

If the document is still on the computer's hard drive and the next user finds it, clear indicators let the finder know whether the document is still current or can be safely discarded.

Include a Cover Sheet

Be sure to prepare a cover sheet or cover e-mail to go with the draft. In it, be very clear about what you want your reviewers to do with the draft, if there are any points you need them to pay special attention to (or disregard), and be sure to give a specific date and time on which their review comments must be returned to you. Some writers also invite reviewers to call if they have a question or want to clarify a point. Most reviewers would anyway, without the invitation, but it doesn't hurt to offer it.

Be Firm About Return Dates

Even if you know you'll accept review comments after the response date and time you specify in your cover letter, keep that knowledge to yourself. The computer industry is driven by perceived deadlines as much as by actual ones, and let's face it, everyone is looking for deadlines they can ignore. Reviewers might not always return their comments on time, but by giving them a clear target, you give yourself room to decide whether to incorporate feedback received "too late."

This last point is particularly important and one that gives tech writers the most difficulty in standing firm about. On one hand, you want feedback from everyone on your review list, or why are they on it? On the other hand, you have your own deadlines to meet and you need reviewers to respect that. Finding a balance is something you have to do on a case-by-case basis.

Pros Know

At a tabletop review, try having one writer act as the moderator while the document's author takes the notes. This method allows you to concentrate on collecting the information with no distractions. So that you can write only new information that arises and not try to take down every last detail, ask to keep (or make copies of) the annotated drafts the participants bring to the session.

Conducting the Review

There are only so many ways to collect feedback from a group of people. The technologies keep getting more gee-whiz, but the goal remains the same: consensus among a group of people about a document's content, completeness, and organization. The point at which consensus is reached happens at various places in the feedback process, depending on what review method you use.

The Tabletop Review

This method, although one of the most effective, also is the one that feels most like a marathon. In a tabletop review, all the reviewers gather in a conference room around a table (hence "tabletop") and together go through the document page by page, paragraph by paragraph, even word by word if necessary. Everyone's feedback is shared with the group and any disagreements or clarifications can be handled as they come up. You, as the writer and review facilitator, keep the focus on the document and take note of what changes need to be made.

Facilitating a tabletop review is not for the faint of heart. Side issues inevitably will arise and you have to be able to (tactfully!) keep the discussion focused on the document at hand. Lengthy and sometimes heated dialogs occur between reviewers at times, and as facilitator you must have the confidence, calmness, and plain chutzpah to take back control of the proceedings.

To:	Engineering reviewers
	Tech support reviewer
	Customer support reviewer
	Marketing reviewer
From:	Kim Reiter
Subject:	Tabletop review of *E-Buy&Sell 2.0 User's Guide*
	June 8, 9:00 - 12:00
	2nd floor north conference room

This is an invitation to a tabletop review for the first draft of *E-Buy&Sell 2.0 User's Guide* on **June 8 from 9:00 - 12:00 in the 2nd floor north conference room**

The document draft copy is available on
I:\Documentation\UserGuide\2UG.pdf.

If you can't be at the review, please mark up the draft copy with comments and return it to me by June 11.

If you can't be at the tabletop for the entire session, please make sure you attend for the sections that concern you:

    9:00 - 10:00: Chapters 1 and 2 (Joe, Raj, Sally, Tania, Ray)

    10:00 - 11:00: Chapter 3 (Bob, Evelyn, Hoang)

    11:00 - 12:00: Chapter 4 and appendices (Juan, Taro, Alan)

Thanks,

Kim

kim@ebuyandsell.com

Extension: 5555

The invitation to your tabletop review should clearly state when and where the review is and what the reviewers can expect to do while there.

You might have to plan several hours and possibly more than one day for a tabletop review, depending on the length and complexity of your document and the number of reviewers involved. However, many companies expect meetings to last an hour, at most two. People on tight schedules will not agree to commit themselves for as long as you'd like to have them there.

One solution is to arrange to review the document in chunks. For example, you might plan to go over the first three chapters in the first hour, the second three in the second hour, and so on. Send out review notices only to the people who need to be there at any given time, and tell them how long the entire review is scheduled to last.

Schedule the meeting room for the entire period of time you will need it. Interested reviewers can stay through all or part of the session. Reviewers also can drop in without feeling the pressure of having to be there for a several-hour stretch. And sometimes people who don't expect to stay will become involved and want to continue.

If your review is going to last more than three hours, it helps if you can have lunch or a snack brought in, to keep the group together and momentum rolling. Ask if your

131

company will provide something, preferably not sweet. This isn't you playing Mom or Dad: It's a serious tactic for keeping reviewers on task. When blood sugar gets low, tempers heat up and thinking gets fuzzy—and you can be sure some of your reviewers got their last "meal" from the vending machine down the hall.

Dodging Bullets

If you arrange to have food brought in during a review, avoid sweets (a "crash" always follows the "sugar rush") and heavy, fatty foods such as pizza. Fruit-and-veggie trays with dips, alone or accompanied by bagels and cream cheese are always popular. If you like to bake, you can bring something yourself, but don't feel you have to make a habit of this. A surprising number of tech writers think they have to bribe developers to review their documents. Remember, it's part of a SME's job to participate in the review cycle, and you should make food only if it's something you enjoy doing and if you are certain it won't hurt your professional image.

If you can't finish the review in one session, state clearly when and where the review will resume, and actively ask for a clear commitment from each reviewer. Don't assume someone will come next time just because he or she is nodding in agreement now.

The strength of a tabletop review is that you get all the involved parties together in a room and hammer things out once and for all. The weakness of this type of review is the challenge of scheduling such a large chunk of time (or more than one chunk of time) for so many busy people. Logistics can be a problem, too, when people are spread across town or across time zones. Teleconferencing works moderately well, but there's nothing like having everyone there in person.

The Distributed Copies Review

This type of review works best when your confidence about the document is very high, your list of reviewers is short, you have at least a week for the review to take place, and you know the reviewers will turn their copies around in the timeframe you need.

To do a distributed copies review, prepare identical *hard copies* of the draft. Staple the cover sheet to the top of each one saying what you want the reviewers to do and when you need the marked-up copies back. Explicitly state that you want them to write on the draft copy and send it back to you. It's a good idea to highlight the name of each reviewer in the reviewer list on the cover letter; this ensures there's a copy for everyone and identifies who the feedback is from when you get each one back.

To:	Engineering reviewers Tech support reviewer Customer support reviewer Marketing reviewer
From:	Kim Reiter
Subject:	First draft of *E-Buy&Sell 2.0 User's Guide*, 6/22/2001

I've attached a copy of the second draft of the *E-Buy&Sell 2.0 User's Guide*. Please return the hard copy to me with comments by end of day, June 27.

I put revision bars on all the sections that have changed.

Engineering: Please confirm the hardware/software requirements on page 7.

Tech support: I left a placeholder in Appendix A for the information you promised me.

This document goes to press June 29, so this is the last chance for changes. Call me if you have any questions or concerns.

Thanks,

Kim

kim@ebuyandsell.com

Extension: 5555

A cover sheet for a document draft adds a professional touch and lets the reviewers know what your expectations are.

If you wish, you can offer the option of taking their feedback over the phone, which you'll note on your master copy, but think carefully before you say you'll do this. It's always better to have something tangible with comments written on it for your file. If one (or more) of your reviewers has a history of changing horses in midstream, you're better off asking for written responses.

Then, deliver the copies to the reviewers. If you have time, delivering them personally is good—if you're lucky, you can put it directly into the reviewer's hands or at least exchange greetings and say, "Here's that draft I told you would be coming." If you don't have time for hand delivery, use whatever method will put the copies into the reviewers' hands most quickly—interoffice mail, Federal Express, and so forth.

Make sure all reviewers will have at least the same minimum amount of time to review the draft—for example, if everyone needs at least three days, make sure you allow for transport time to and from your reviewers on both sides of that three-day window. Another option is for reviewers to mark their copies and then fax their

Tech Talk

A **hard copy** is a document actually printed on paper, whether from a laser printer or an offset printing company.

Dodging Bullets

Let's face it—reviewers just don't like to receive a huge thick document, all of which needs to be read and checked. To get a better response to your review requests, send out the smallest chunk of the document you can to each reviewer. It might take extra time up front to break up your document into these SME-specific chunks, but you will get a much better return on your investment—a lot more people read small sections. And you might save some trees, too.

feedback to you, but this backfires if a reviewer's handwriting is small or hard to read, or if comments written in a margin extend past the area the fax machine reads. It's worth trying, but be warned.

If you are lucky enough to be able to give your reviewers a longer review window, you might want to call or e-mail them with a friendly reminder a few days before their feedback is due on your desk. Things have a way of getting lost on people's desks, and the longer they are there, the deeper they get buried. Most people will appreciate a reminder so something they agreed to do doesn't slip through the cracks.

If all the review copies don't come back by the date and time you specified, you'll have to decide what to do based on your particular circumstances. Ask yourself these questions: How important is the reviewer whose feedback you don't have? Will it set a counterproductive precedent if you don't follow up? Your answers will help you decide on a course of action. And don't forget about your boss or supervisor—he or she probably needs to be in on your decision-making process and might even be able to elicit the response you need from the wayward reviewer.

The strength of the distributed copies review is that reviewers can go through the document at their own pace and at different times and locations—nobody has to coordinate with anybody else. You also receive written responses, which go straight into your file for the document, in case there's a question later about who wanted what change and when.

The main weakness of the distributed copies review is that it depends on reviewer follow-through—he or she has to independently make time to do the review, write the changes on the copy, and send it back to you on time. This is expecting a lot of someone who has six other people actively tugging at a sleeve. This review method also takes more time—not in actual review hours but in calendar days—and in the computer industry, time is something you seldom have.

The Electronic Review

The electronic review is becoming more and more popular—and it's no wonder. Many people now can type faster than they can write, and typing their comments directly into a *soft copy* saves precious time. Word processing and DTP programs also usually have features that let you track changes made to a document so you, the writer, can readily spot exactly where and how a reviewer has changed the draft.

Document files attached to an e-mail cover letter can be sent around the world in a matter of minutes, compared to paper documents that can take days to get from one place to another. There's also no bulky paper to clutter up a desk or briefcase, and it's lots harder to lose a file than a collection of paper pages. What's not to love?

Well, some people just need to see it on paper and write on it with a pen or pencil to get their brains going. But those people can easily print it out and review it that way. They may or may not type in their comments afterward, but sending a soft copy and getting paper back is still better for you than getting nothing back. Go with the flow—at least you got the feedback.

There are two ways to conduct an electronic review. The method you choose depends on things such as how cooperative the reviewers are and how much time you have.

The collaborative electronic review puts the document in the hands of one reviewer at a time. He or she adds feedback and sends the file on to the next reviewer on the list. This gives each reviewer the benefit of seeing what others have already addressed and eliminates duplication of effort. If there are questions or disagreements, each reviewer knows exactly who to speak with and often can iron things out before the file makes its way back to you.

Tech Talk

A **soft copy** or **electronic copy** is a document that is read from the computer screen.

Pros Know

Make it very clear exactly what you need reviewers' help with: content. Some reviewers, perhaps frustrated writers, might spend time and attention on aspects of the document such as punctuation, style, and grammar. Or a marketing reviewer might not like the format or tone. But that's not his or her area of expertise or involvement, nor is it input you can use. Making this expectation clear at the outset helps everyone make the best use of time and energy.

The other method of electronic review is pretty much like the distributed copies review, with the same attendant issues—the only difference is that you distribute a document file instead of a printed copy, and the reviewer feedback is typed in instead of written by hand. Depending on your e-mail reader, you might have to be careful about file names when you start receiving many copies of the same document back from reviewers.

There are variations on both of these methods. Perhaps a manager is your primary reviewer, but he or she circulates the file to others in the same department before consolidating input and returning the file to you. Or you might send a file to a short list of reviewers, then sit down individually with each reviewer while he or she gives you verbal feedback.

The value of an electronic review depends a great deal on how thorough and articulate the reviewers are. A file that returns with "Looks okay to me" typed on it really isn't much help; nor is one that is peppered with cryptic remarks like "What's this?" or "Doesn't work this way." You might have to try it once or twice before you can tell if this method will work at your company. Bear in mind that if the feedback isn't clear or raises more questions than it answers, you'll need extra time to unravel knots and account for loose ends. Your deadline could be at risk.

Who's On First, What's On Second

Now that you've heard back from all your reviewers, it's time to get down to the task of consolidating the feedback and deciding how best to incorporate it into your document.

Consolidating Reviewer Feedback

The way you approach this part of the process depends on your review method. If you conducted a tabletop review, you probably already have all your notes gathered in one place because you wrote them down during the review meetings.

If you used some form of electronic review, the feedback might all be in the file but it might take some sorting through to get to the bottom of it. If dialogs between reviewers were captured in the document, for example, make sure you follow them to the end and incorporate only the final change they've requested. If you have multiple files, each with a reviewer's feedback in it, you might want to use your word processor's or DTP's feature for comparing documents.

If you sent out printed review copies and now have your stack back with the reviewer comments written on them, you might want to collect all the notes on a single, master printed copy. This gives you a single source to refer to when you are at the keyboard, and it can go into your file in case there's a question later about when a change was requested or made.

One good way to incorporate reviewers' comments is to systematically go through each review copy and type the new information into your master file, one item at a time. As you incorporate each comment, cross it out in the reviewer copy.

Coffee Break

Scott Adams, creator of the *Dilbert* cartoons, created Tina the "brittle" technical writer. Which came first—the stereotype of technical writers as hypersensitive, or technical writers who really behave this way? Scott Adams tells all in this mini-interview he granted to us:

Who (or what) inspired Tina?

I worked with a tech writer who believed that all forms of communication were veiled insults against her.

What kind of mail regarding Tina do you get from tech writers?

Some hate it for typecasting them. Most like the attention given to their profession.

Is there any career path for Tina or is she stuck, too?

Tech writers are considered glorified secretaries in many companies. They are doomed unless they change companies and professions at the same time.

Generally it will not be difficult to know what changes to include and where they need to go in your document. Many times every reviewer will have picked up on the same few items. As you build upon your original file, you are likely to notice that the same issues come up again and again.

It's also common for each reviewer to provide more feedback in the part that falls in his or her area of greatest expertise, with lighter feedback elsewhere. Ideally you will have solid feedback from at least one reviewer for every major part in your document.

As you're analyzing the feedback, you might find yourself feeling unsure exactly what to do with some of it. In these cases, it helps to remember one fundamental truth: You are working as a writer because of your expertise with writing. So have confidence in your own ability to choose the right words and to say things in the way appropriate for your audience. What you need from reviewers is feedback about the content of the document—whether explanations are complete and accurate, whether steps appear in the correct order, whether diagrams and tables depict their information correctly. These are the kinds of feedback to focus on in whatever you get back from your reviewers.

Resolving Conflicting Feedback

Sometimes it happens that two (or more) reviewers make statements that flatly contradict one another. First try to resolve this with phone calls or e-mails to the reviewers. Make sure you're interpreting their feedback as they intended and clarify any possible points of confusion. If this doesn't clear things up, the only thing to do is to get the reviewers to talk with each other and reach a consensus.

This might be something that can happen casually—perhaps one says he or she will call the others and let you know what they all decide. At other times it might need to happen more formally, such as in a meeting around a conference table. In this situation, it's best to set the ball rolling by recapping what you understand to be the point of disagreement, then sit back and let the reviewers work it out.

In such a situation, your position is neutral. When it seems they've reached agreement, be sure to restate what you've heard so you'll know you've understood and captured it correctly. Depending on your company's climate and culture, it also might be worthwhile summarizing the issue's resolution in a memo or an e-mail—one copy to each reviewer involved, and one copy for your files.

> **Pros Know**
>
> "Just getting a review can be like pulling teeth. Assuming you've managed to bribe, needle, cajole, or nag someone into giving your draft a review, the next step is to put on your thick skin. You'll hear things you don't want to hear about how you phrased something in a confusing way or bungled a procedure .... Take it in stride and appreciate their effort—sometimes they're right!"
>
> —Betsy Pfister, Technical Communications Director, Red Hen Systems, Inc., http://www.redhensystems.com

A somewhat delicate point to keep in mind is the relative "clout" of each of the reviewers involved in a disagreement. If they can't resolve things in a way that satisfies both or all, you'll have to decide how to handle the issue in your document. A good rule of thumb is to accept the higher ranking reviewer's position, even if you don't agree with it yourself. It's up to the developer to take up the issue with his or her own supervisor at that point. You indeed might come back to it and change it later, but the issue needs to be pursued through other channels before that can happen.

If you feel strongly that the higher ranking reviewer's position isn't the one that should be implemented, it's worthwhile to talk with your own supervisor about the best way to address the issue in the document. He or she might say to go with it anyway, or to hold off until the issue is resolved. Either way, it's now a shared burden instead of being all on you.

Once More, with Feeling

After you've collected all the feedback, resolved reviewer conflicts, and incorporated the changes and additions into your document you're ready to … do it all over again. Yes, we're serious. But with this round, things will be a little different.

Everyone has seen the document before, so they know what to expect. You'll want them to focus on the changes and additions rather than reading the whole thing over again. Make sure these aspects are clearly marked with revision bars or at least a page of notes explaining what's different.

Reviewers don't want to see the same document again and again. Their schedules are tight, and however much they'd like to help, when they read a document once, most of them feel they've done their job. Your rate of return on a second or—horrors!—third review copy is likely to be much smaller than the first, when the information is new and fresh. And if a reviewer had little to say the first time he or she read the draft, consider whether it's even worth sending another copy.

How many repetitions of this review cycle do you need? The theoretical answer to that question is "however many it takes to get the content right." But the practical answer is "not more than one or two." If you've done your homework, gathered all the information, organized it lucidly, talked with the right people, and solicited and incorporated feedback from knowledgeable reviewers, you shouldn't need more than a few review cycles, and there's rarely time for more than two, anyway.

But things don't always go like clockwork: The product might change radically; the people involved in a project could change; an internal process might change; even a company's overall business objectives or market positioning can change. We've worked with one telecommunications company that has, without fail, experienced a reorganization at the department level or higher during every project we've had with that client over a period of eight years!

At some point—whether for the second review or the fifth—you'll feel confident that this review is the last. It's time to get final approvals and sign-offs. Send a sign-off sheet with the printed copies, or circulate a separate sign-off sheet to the people authorized to give your project the official green light. Make sure it comes back to you with all the required signatures in place. If your company uses electronic signatures, you might simply need to send an e-mail; they should reply to it with their signature file attached. How you word the sign-off sheet is up to you, but it needs to do the following:

➤ Clearly identify the document by name, product, version, and date.

➤ Include an explicit statement of acceptance or approval and whether it is a final draft or an interim draft (this may or may not include the words "with changes as marked").

➤ Provide space for the signature and spell out the name and title of the signer.

➤ Provide a labeled space for the date of signing.

Document Sign-Off

Document title: *E-Buy&Sell 2.0 User's Guide*

Writer: Kim Reiter

Technical adviser: Tania S.

I have reviewed   [√] All of this document

                 [   ] Sections [   ] of this document

	Approve	Approval contingent upon the following changes:
Document is complete		*Needs procedure added to Chapter 3*
Content is technically accurate		*Corrections on pp. 7, 8, Chapter 3, 32-34*
Graphics are correct	√	
Level of detail is appropriate	√	

Signature: **Tania S.**

Date: *6/11*

A sign-off sheet can come in handy if disputes arise later.

Congratulations! You've shepherded your document successfully through the review process. Now you can rest assured that the content is solid, accurate, and complete.

The Least You Need to Know

➤ Choose reviewers who already know the information and can give you meaningful feedback.

➤ Spell out for reviewers what you want them to do and how much time they have.

➤ Prepare the review draft carefully to minimize distractions such as typos, and make sure it looks as finished as possible.

➤ Bring reviewers together to resolve conflicting feedback; summarize what they agree on.

➤ Count on more than one review cycle, but fewer than four.

Chapter 13

Front and Back Matter (They Sure Do!)

In This Chapter

➤ Starting your document on the right foot

➤ What comes after the last chapter

➤ The importance of the index

In establishing a relationship with your reader, it's important to get off on the right foot. First impressions matter here just as elsewhere. Luckily, with a document you have great control of the kind of first impression you create.

Similarly, when you take your leave at the end of a document it needs to be a graceful exit—no stopping in mid-sentence. In this chapter we'll walk you through front and back matter so you can see why they *do* matter.

About Front Matter

The first few pages of your document usually consist of a page with copyright and other information such as trademark acknowledgments, a table of contents, sometimes a list of figures or illustrations, and nearly always an introduction. These pages are important because they orient the user to the rest of the document, providing a first big-picture look at what the document contains. They set the reader's expectations for the rest of the document.

Copyright Information

The *copyright* statement (called a *notice*) usually appears on the first page of a printed document, even before the table of contents or the first page of text. In online documents, expect it to appear on the first page a viewer sees and possibly on every copyrighted page the viewer can link to from that first page. The copyright notice is important because it establishes ownership of the document's intellectual property.

The accepted format for a copyright statement is the copyright symbol (©) followed by the year, the name of the copyright owner, and the words "All rights reserved." It also might include a disclaimer spelling out what the copyright notice means. In a technical document that will be something along the lines of, "No part of this publication may be reproduced, stored in or introduced into a retrieval system, or transmitted, in any form or by any means (electronic, mechanical, photographic, audio, or otherwise) without prior written permission of [Company Name]."

Tech Talk

Copyright is the exclusive ownership of a literary work, a musical work, or a work of art. The copyright owner retains the right to make use of the work, which is protected by law for a specific period of time.

When a document is created, the copyright year is easy to know: It's the current year. But it isn't always quite so clear-cut for later editions when substantial portions of the document might be changed, or when major sections are added.

Here's the guidance offered by *Microsoft Manual of Style for Technical Publications, Second Edition:* "If 85 percent or more of a piece is new at reprinting, it is considered a new work, and the copyright should list only the current year at the time of the reprinting. If less than 85 percent of a piece is new at reprinting, it is considered a derivative work, and the copyright should list both the original year of printing and the current year at the time of reprinting." For example, the copyright notice for a document from the fictitious company We're Widgets, Inc., created in 1998 and reprinted in 2000 with less than 85 percent changed, would be "© 1998, 2000 We're Widgets, Inc."

But we're not lawyers (thank goodness) and the advice in this chapter is only a recommendation. As with all such legal issues, always check with your company's legal department or get other professional legal advice about copyright and other notices that your document needs to contain, such as trademark acknowledgments.

Table of Contents

This is the core of your front matter. The table of contents not only introduces the reader to the document; it shows how information in the document is arranged, what is and isn't included, and even gives broad hints about tasks or procedures involved in using a product. That's a lot for some headings and page numbers!

Because every document needs a table of contents, and what goes into a table of contents is derived directly from the document's text, a feature that automatically generates a table of contents is standard on word processing programs and desktop publishing systems. Some do a better job than others, especially in creating one across a multidocument book. Try not to do too much editing or tweaking of a generated table of contents—whatever you do once, you'll have to do every time. It's better to see if you can fix the problem in the text itself.

For online documents, obviously you don't need page numbers but do use some visual cues to let users know that this is the table of contents. Create a table of contents that links to the actual place in the document; users don't like seeing exactly where they want to go but being told, "You can't get there from here." Make sure they can (Acrobat PDF files create linked tables of contents with their bookmark feature). If your online documents are created in HTML, test the links to make sure headings take users to the right places, and that the heading at the top of each target page is exactly the same as the heading a user clicked to get there.

Pros Know

Because information is so volatile and documents need constant updating, make this process easier by numbering chapters and sections independently. Use a two-part page number: the first number for the chapter or main section, the second number for the page within that section. If you start each new chapter with the page number "1," you can replace chapters or parts of chapters without affecting the rest of the document.

Most tables of contents have no more than three levels of headings (four if there also are major document divisions to identify). Usually the highest-level heading is the chapter title. Under that are the headings for the main chapter sections, and under them, headings for the subsections.

Tables of contents are good mirrors in which you can really see your document's structure. Have you been consistent from one chapter to the next? If not, take a look at how you can make chapter titles more parallel. Have you used more than three levels of headings in any chapters? If so, make sure the third-level headings give a clear indication there are multiple subdivisions by using a plural. For example, a third-level heading might be "Setting Optional Parameters." The fourth-level headings the reader would find in the text (but not the table of contents) would be "Setting Parameter A," "Setting Parameter B," and so on.

If you've followed clear writing guidelines (see Chapter 19, "Writing Clearly"), your headings use the gerund form of verbs in the text and naturally will in the table of contents. This is important in guiding readers to the right section because they probably know what they want to accomplish even if they don't know exactly what feature to use.

Coffee Break

An element found in other books (such as this one) but not in technical documents is the foreword. Why not? It's a short introductory statement usually written by someone other than the author that provides commentary about the book. Does any of that sound useful to someone referring to a technical document? You're right—it's simply not needed.

Make sure shifts in heading levels are obvious in the table of contents by using strong visual cues such as indents and clear differences such as type size, bolding, and italics. Separate major sections with plenty of white space, too, so the reader can tell at a glance what headings belong together. The structure of the table of contents should still be evident when you view the page from 3 feet (1 meter) away.

If your document is figure intensive, you might want to follow the table of contents with a list of figures—or maybe two. Why two? Depending on the nature of your document's illustrations, a user might want to find a particular illustration by the chapter or section where it occurs, or by the illustration's subject (indicated by its title or caption). For online documents, you might be able to let the user sort the illustrations by either placement or subject. For printed documents, include two lists and make it obvious what the ordering principle is in each one (use the document location or illustration title).

Introduction

"Hello, nice to meet you, my name is ..." That's what your introduction needs to convey to the reader. Here are the points your introduction should cover, in approximately this order:

➤ What product this document relates to, giving a specific release or version identifier

➤ Who this document is written for—this includes others who also might benefit from reading it in addition to the intended audience

➤ A high-level overview of what information the document contains, usually in the form of an annotated table of contents that briefly describes what's in each chapter

➤ Conventions used in the document, such as different typefaces (see the following section, "Typographical Conventions")

You also might want to include an overview of new features or radical design changes. These points are especially important to experienced users who might be expecting this product to work just like the last version. Letting them know up front (literally) to expect these changes helps reduce the shock when they actually encounter them in action.

The introduction also is where you set the tone for the document. The writing here needs to be especially crisp and clean, and perhaps even a bit more cordial than the rest of the document. After all, you smile, don't you, the first time you meet someone?

Typographical Conventions

It's common to use different typefaces, italics, and bolding to identify code samples, error message samples, command and environment variables, user input, system output, and other vital kinds of special information in your document. The typographical conventions section, usually part of the introduction, is the key your readers use to unlock the conventions that "encode" meaning into the text.

The most efficient way to convey typographical conventions is with a simple box *matrix* like the one shown in the following figure. On the left, name the convention (a font name, or a variation such as bold or italics). In the next column, use focused phrases to describe what the convention is used for. Then, on the right, give an example of the convention in use.

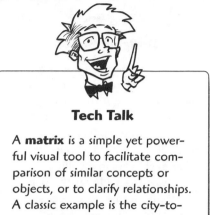

Tech Talk

A **matrix** is a simple yet powerful visual tool to facilitate comparison of similar concepts or objects, or to clarify relationships. A classic example is the city–to–city mileage matrix in your road atlas. Another is a comparison chart from *Consumer Reports* magazine.

Convention	Meaning	Example
Italics	Key words, such as terms that are defined in the text	If the transaction is authorized, a capture takes place.
	Titles of books and documents	For more information, refer to *E-Buy&Sell Administrator's Guide*.
	Emphasis	Under no circumstances reveal your password.
`Courier bold`	User input	To change to the E-Buy&Sell configuration directory, type: `$ cd $EBBASE/EBuySell/Works/config`

`Courier`	text file content	`<HTML>`
		
		`<TITLE>Ebuy$Sell Home Page</TITLE>`

	machines messages and code samples	`term = new Terminal ();`
	file names and URLs	See `www.ebuyandsell.com` for information about EBut&Sell products
▲	Indicates a caution. Caution means there is a possible danger to equipment. Look for this icon in the left column.	

Here are some of the typographical conventions you might put in the beginning of your document.

It's essential for your samples to match the convention they're illustrating! Readers find it highly confusing (not to mention irritating) when, for example, the convention for user input is Arial italic but the sample is done in Times Roman italic. They don't know which one to believe. Avoid creating such "do as I say, not as I do" situations.

Many documentation departments are phasing out the typographical conventions section, so this may or may not be part of your company's standard. The rationale for eliminating this section, long assumed to be an essential part of the introduction, is that if the tech writer is consistent in the use of the typographical conventions, there's no need to spell them out. After all, the user understands consistent use the same way you understand which lines in this book are headings and which are body text.

It probably is safe to not use a typographical conventions section as long as you are consistent in your use of conventions. However, if your document contains warnings and cautions, it is a good idea to explain them to your user. In this section, tell the user what typeface or icons differentiate notes, warnings, and cautions. (We explain the difference between warnings and cautions in Chapter 18, "Style Guides: Not Just a Fashion Statement.")

Dodging Bullets

A table can be a good way to condense reference material for an appendix. But think carefully about how to structure it, or it could be trouble. Put the most fixed list in a row and the list that might vary in a column. That way the table fits the page width, but can extend over several pages if needed and is adjusted easily for revisions.

About Back Matter

You might think that because back matter is, well, at the back, that it doesn't matter very much. But you'd be wrong. The back of a technical document is where users expect to find several types of important information; the common characteristic of all these items is that their use is particularly infrequent or to some degree optional:

➤ Detailed, in-depth reference information about the product (such as parameter tables, field definitions, or variable values)

➤ Specialized information for optional features or enhancements that not every user has (but that some probably paid extra for and want to get the most out of)

➤ Insights for how to make the product work optimally with another product (or more than one)

➤ A glossary of terms specific to the product (which also might be essential to understanding the document's concepts and instructions)

➤ That unglamorous but essential feature, the index

Appendixes

This broad category of back matter covers anything that isn't a glossary or an index. Appendix titles usually are identified by letters instead of the numbered titles used in chapters. This provides a sense of order and connection among small bodies of information—the individual appendixes—that otherwise might not relate to each other at all.

Because they are individual bodies of information and are likely to change frequently as well as independently of each other, number the pages in each appendix separately.

Glossaries

The key thing with a glossary is to strike a balance: Define terms that are specific to the product or the document, but don't overload the user with too many entries. Rather than a "when in doubt, spell it out" approach, be selective in the terms you include. For instance, perhaps a term isn't used in everyday conversation, but is well known within your industry. Leave that one out. If, on the other hand, a common or vague term is used in a highly specific way in relation to your product, by all means list it.

Put yourself in the new user's place as much as you can. If you simply are so accustomed to using and talking about the product that it seems like second nature to you, find a stand-in new user. Ask this person to look at the document and glossary for you and point out other terms to include or entries that need clarifying.

One more thing: Even if your glossary is short, split up the entries by letter rather than simply running all the entries together into one long list. This simple step can significantly reduce the user's search time, allowing him or her to focus on understanding the meaning of a term, instead of just finding it.

Dodging Bullets

Don't let appendixes be the dumping grounds for information you don't know what else to do with. If you're not sure how a body of information relates to the rest of your document, maybe it doesn't. Sometimes information put into an appendix really needs to be given a life of its own as an independent document, or to be included in a different document altogether.

Coffee Break

Ever wonder about the history of a word—where it came from, when it was first used, how its meaning has changed over time? Take a look at the *Oxford English Dictionary*. A full-sized set is twelve thick volumes, but it's also published in a compact edition: two *very* thick volumes, four "regular" pages microprinted on each thin page, and comes with its own magnifying glass (thank goodness).

The Index

One of the most important pieces of back matter is the index. This is where readers turn even before looking at the table of contents. When people turn to an index, they need to find a specific piece of information, usually in a hurry.

Here's an image that might help you: In the movie *Air Force One,* there's a scene in which the U.S. president, played by Harrison Ford, is holed up in the baggage compartment of the presidential plane while terrorists hold the plane's other occupants captive. Luckily for him, he's found a cell phone in someone's bag—but doesn't know how to use it. Frantically he leafs through the accompanying set of instructions, while the lives of his loved ones and staff hang in the balance. That's the user image to hold in your mind when you set about creating your index.

Does your document really need an index? Well, is it more than four pages? Then it needs an index—and a good one. A poorly done index is worse than none. Be sure to see Chapter 14, "Indexing," for how to index your document well.

You may create the front and back matter as the last steps in developing your document, but they are far from being least important. Like a painter who knows that the background of a painting can be almost as important as the subject, give them the time and attention they deserve so that they, like the rest of the document, are complete, accurate, and usable.

The Least You Need to Know

➤ Check with your company's legal department or other legal advisor to be sure your document's copyright notice is complete.

➤ Limit your table of contents to no more than three levels of headings.

➤ Use the introduction to set user expectations for the document.

➤ Make sure each appendix relates to the rest of the document in some way—if it doesn't, take it out.

➤ Split up glossary entries by letter rather than running them together into one long list.

➤ If your document is more than four pages, do an index—and do it well (a poorly done index is worse than none).

Chapter 14

Indexing

In This Chapter

➤ Why users love indexes

➤ Why creating and maintaining an index is nobody's favorite task

➤ What makes a good index

An index is an essential part of a technical document—for both print documentation and Web documentation. Imagine driving around in a strange city without a map, trying to find your friend's house but having no idea what street he or she lives on. You drive until you see a house that looks as if it might belong to your friend—the color looks right, and is that the right car in the driveway? All you can do is stop and ask, and too bad for you if there's nobody home. That's what it's like using a document that has no index.

If you release a customer document without an index, it is almost guaranteed that the users will complain, and with good reason. But a poorly done index can be worse than none at all—would you rather have no map, or one that looks okay but in fact is wrong? Indexing must be done well if it's done at all.

It takes time and practice to do indexing well, and we can't solve all your indexing problems in one short chapter. But we can give you some guidance on what makes a good index and how you can improve the indexes you create for your documents. That's what we do in this chapter.

Why an Index?

Okay, all you tech writers, let's face it: Hard as it is to accept, the truth is that users seldom read an entire document. Oh, yes, there are always a few people who obsessively read an entire user guide before ever touching the product, or people who take product books with them to read while they travel or are in a hotel room. And we love them for it—but they're in the minority. Most users of any technology refer to the documentation (print *or* online) only to fulfill a specific need. And when they do that, they want that piece of information right away. No, not right away—*now*.

Although your users probably will glance at the table of contents to see what type of information is covered in a given document, the index is where they'll turn when they really want to find something, and find it fast. It is a pointer to pages that contain relevant information.

Index entries, along with table of contents entries, are called "entry points." They are points, like gateways, where you can enter the body of information stored in the whole document. When people are in a hurry, they don't want to take the scenic route—they want a gateway that's a shortcut directly to what they want to know.

It Helps You Find Your Way

Using a document without an index is frustrating, to say the least. Maybe the table of contents points you to a general area of a document, but after that, you must read through an entire section to find the information you seek—if it's even where you're looking. And if you don't know or can't remember the name of what you're looking for, a table of contents doesn't help at all.

Dodging Bullets

Don't make the mistake of thinking you can get away with omitting the index for your print or Web-based documentation. One of the biggest user complaints about documentation is lack of an index, or an index that's poorly done.

Put yourself in the reader's place for a moment. If you know there's a UNIX command that allows you to apply a fixed set of editing changes to a sequence of files, and you need it right now but you just can't remember the name, what do you do? Skim through a 400-page document hoping to get lucky until you run out of patience? And how much time will you spend before you find it, or waste looking if you don't find it? With a good index, finding what you want is quick and clean: You look up "editing" and find that one of the commands is "sed." Ah yes, that was it!

But Doesn't the Table of Contents Do That?

A table of contents isn't the same as an index for several reasons. First, a table of contents can't (and shouldn't) contain the level of detail an index does. Here is a part of a table of contents for a photo-retouching program:

Mask Tools

Manual Mask

Color Sample Mask

Intensity Mask

Gradient Mask

If you don't already know what each of those terms means, the table of contents won't help you. Some of the terminology in this sample is, in fact, specific to this program and is not used industry-wide. So even an experienced pre-press specialist might not know what to look for. To further complicate things, an experienced user, new to this software but not the processes, will search for terminology he or she is already familiar with. What happens when it's not in the table of contents? That experienced user turns to the index, and finds these entries for the *same* pages listed in the table of contents:

color correcting, with a mask

drop shadow, *see* shadow

merging two images together

shadow, creating on an image

silhouette, creating with a mask

And so on. Also notice the two entries "drop shadow" and "shadow." Those entries actually mean the same thing and point to the same page. This method acknowledges that people use different terms for the same thing, and means that no matter which term the user looks up, he or she can find the desired information.

Why Everyone Hates Indexing

It's not uncommon to hear tech writers groan when told they have to index their document. Many writers hate to index. They don't like to go back after the document is finished and figure out what should be indexed. They find indexing to be tedious

Pros Know

Wouldn't know where to find an indexer? There are national organizations devoted to indexing. Go to their Web sites at

American Society of Indexers: http://www.asindexing.org/

Indexing and Abstracting Society of Canada: http://tornade.ere. umontreal.ca/~turner/iasc/ home.html

Tech Talk

An **index marker** is the symbol generated by a word processing or desktop publishing program that indicates where an index entry is. Index entries are hidden in a normal page view, and can be seen only by searching with a special command.

and time-consuming and it always has to be generated several times before it's right. Worse yet, they don't understand what makes a good index in the first place. Okay, there are some tech writers who enjoy indexing and are good at it (although they might hesitate to admit this!). But the ones who dislike it know that there are people who make their livings as professional indexers, and they can do your index quickly, and do it right.

But hiring a professional indexer is not an option most tech writers have; there's rarely the time or the money for it. No, indexing is usually the responsibility of the tech writer.

In addition to being time consuming and thought intensive, indexes are difficult to maintain. Why? Anytime you update the document, you have to update the index entries, too, and people manage to forget that. Desktop publishing and word processing programs typically hide the index entries, so there's rarely an easy way to find the *index markers*. Out of sight, out of mind. That is, until a user wants to find something.

Another Document's Markers

Another indexing problem crops up when you've pasted in text or structures from other documents. Let's say Joe writes an administrator's guide that follows the correct structure, and Kim sensibly copies the whole chapter into her new document. She keeps all the heads, because they are the same in each document, and also keeps some of the paragraphs that share similar styles, planning to change the product name.

But what Kim may forget is that she's also imported index markers along with everything else. If she doesn't update all the index markers, they're going to contain references to a different document's product.

What Makes a Good Index

Many people connected with documentation misunderstand what an index should be. They might believe that a word search is good enough, so if the documentation is online, they rely on the "Find" command to search for anything the user needs. If the documentation is paper, well, it's anybody's guess how the reader would search for a particular item—leafing through the pages?

All a "Find" command can do is find the word, not take the reader where the word is used with the meaning or context he or she needs. An index entry, for example, showing 30 different page numbers for the term "configure" doesn't help much. The "Find" command also might stop 10 times on a given page, with none of those stops being what the reader wants. Searching for a word won't help a reader locate a concept, task, or phrase that might not be worded that exact way. A good index addresses concepts, ideas, and actions, as well as keywords.

It might have occurred to you that indexing is something computers would be good for. Although there are computer programs that search for keywords and sort them in order, those programs produce only a list of words, usually nouns, with no understanding of whether a word is meaningful in the context of the text. Experimental indexing programs have even been developed to try to duplicate the mental processes of professional indexers; when the automated indexes are compared to human-created ones of the same material, however, the human-created indexes are nearly always better. It just takes a human being to make a good index.

A good human indexer takes a term such as "configure" and explores deeper. The indexer finds out what's being configured and what other actions or results relate to it, and lists those along with the main two or three pages where the concept or action of configuring is addressed.

A human-generated index also can contain words that never show up in the document itself. Is there a shorthand term that's commonly used informally about the product but that wasn't actually used in the text? The indexer can acknowledge this and include it as an index entry with a pointer to either the formal term or directly to the information itself.

Dodging Bullets

Rearranging text or cutting and pasting information can cause you to inadvertently delete or move index entries. Check the index entries in your documents with each update to make sure that the page references don't dead-end, or point to the wrong place.

Tech Talk

A **topic** (also called "**subject**") is the word, phrase, or abbreviation listed in alphabetical order in the index. Together with the reference, topics make up the index entry.

A **reference** (also called "**locator**") is the part of the index entry that takes the user to a place that answers his or her question. It is a page number or set of page numbers, or a pointer to another subject, typically indicated with "*See topic XYZ.*"

Coffee Break

Although it takes a human to make a good index now, you never know what the future holds. In 1968, International Chess Master David Levy bet $3,000 that no chess computer could beat him over the next 10 years. He won his bet, and many predicted that a computer would never effectively play against the best humans.

But on May 11, 1997, chess computer Deep Blue defeated reigning World Champion Garry Kasparov in a classical six-game match. Today, Deep Blue isn't even among the top 10 highest-rated chess computers.

Knowing What the User Wants

You have to be a little bit of a mind reader to create a good index—or at least good at putting yourself in the user's shoes. A good index is one that contains every listing the user *might* think of—and you must imagine what terms the user might use when searching for topics. To create a good index, you must put yourself in the place of the user, maybe even more than when you write the body of the document.

Pros Know

A tech writer who is familiar with both the product and the context of typical job duties in which it's used has a better sense of what terms the user will look for in an index. This writer understands not just *what* the user is looking for, but *why.*

If several users were to look in an index for items relating to Internet Explorer, for example, one might look for "IE," another might look for "Internet," and yet another might look for "Microsoft." And suppose your index has only one listing for it—called "browser"? None of those users will find the entry. You can see why a good index lists all those entries a user might look for, and maybe even a few more (such as "Explorer").

If you have multiple entries all pointing to the same thing, you don't have to list all the same subentries for each. You can list page numbers and subreferences for one or two important terms ("Netscape" and "Internet Explorer," for example) and for an entry like "browser"? A *"see"* reference to the other entries makes everything clear. Make sure if you use a *"see"* reference, however, that the user can find everything there. Choose a main entry where you'll consolidate variations, then point to it consistently. In the following example, a user is directed to the right place whether he or she looks up the term "domains," "fully qualified domain name," or "NT domains." This is a good index in action:

```
domains
    multiple, 57
    NT, 12, 34
    installation checklist for, 9
    single, 56
fully qualified domain name,
    see domains
NT domains,
    see domains
```

How do you know how a reader might look for a term or concept? When a reader starts looking for a piece of information, he or she has a mental image of what that piece of information looks like or what it needs to be like. This is because the reader knows the intellectual context of the information "gap" that the missing piece needs to fill. Along with this, people expect to find "things" (this includes concepts) near other "things" that are related.

Imagine for a moment going to the grocery store in a strange city; maybe you're looking for your favorite bottled water. Water is a drink, so you might look where other drinks are, in the drink aisle. Lots of people like their water cold, so you might also look where other cold things are, in the cooler cases. Some people consider bottled water a "health food" item, so you might also look for it in the health food aisle. And you might find it in any, or all, of these places.

(On the other hand, you might end up playing "outguess the stock boy," and be very thirsty when you finally stumble across what you want after walking the store aisle by aisle. You've just had an object lesson in indexing.)

So, reading the reader's mind means imagining the intellectual context around the gap a piece of information can fill. If you can think of what other information relates to a given word or concept—what your user will think it might be close to—you can create a better index.

Dodging Bullets

When indexing, start with the part your user will already know and point to what he or she won't know. One style guide we used had no entries for "file name," "command," or "field name." We later found the references for those items under the index entry "monospaced font" because they all use a monospaced font. Wrong order: We wanted to know what type conventions to use for those items, and had no way of knowing that in advance.

Descriptive Entries

A good index entry is descriptive. You should be able to tell by looking at it what you'll find on that page and whether it's what you're looking for. Trial and error method even among a few choices wears thin quickly.

No Dead Ends

What's an index dead end? It's an index entry that ultimately doesn't point the reader anywhere. In a good index, every topic contains a reference that ends at a logical place. Usually this is a page number. The page a reference leads to *must* contain information related to the index topic.

The reference also might be a set of page numbers. If an index entry appears on subsequent pages in an unconnected way, list the page numbers separately. In the following example, "IP addresses" entry appears on pages 9, 10, and 11, but not as part of a contiguous section about IP addresses. They are distinct and different references, so the entry would read as follows:

> IP addresses, 5, 9, 10, 11, 15

If the entry appears across pages in a segment of text about that topic, show that the topic is continuous over more than one page. Such an entry would look like this:

> keyboard commands, 17-20

Seeing Is Believing

Sometimes, you use a *"see"* reference instead of a page number to direct the reader. Use this type of reference when, instead of telling the user what page to go to, you want to send the user to another entry. This is a good way to address terms that don't appear in the actual text, and for abbreviations and acronyms:

> VPN, *see* Virtual Private Network

It is best, however, not to send the reader off on a hunt with a *"see"* reference unless the pointed-to entry has some additional information. If "Virtual Private Network" had no subentries, the reader might feel that his or her time was wasted. In a case such as that, put the exact same page number references under both "VPN" and "Virtual Private Network" entries and let them exist independently of each other.

The *"see also"* reference is handy to use after a subentry if there's another entry with additional information that may (or may not) be of interest to the reader. It points to an entry that contains different but related information; it's then up to the reader to decide if the other entry has useful information for his or her particular need. For example …

hosts
    adding
    changing name of
    database for
    definition of
    see also /etc/hosts, Web hosting

Creating an Index

A topic of debate is at what point to create the index. Some books advise that it's far better to create your index while you write than to wait until the end. We believe this is a matter of personal choice and depends greatly on work style. Some people are hopelessly inadequate at placing markers while they write but can race through a completed document and produce a finished, thorough index in no time. Others get bogged down in a completed document but create beautiful, thorough indexes by placing markers as they write. The important thing is to learn the method that works for you and then do it—because indexing isn't an option.

How do you decide if a document needs an index at all? The guideline we follow is that all customer documents containing more than four pages plus a table of contents should be indexed.

Manual vs. Computer Indexing

Now that you know how important an index is, do you know the best way to create one? Indexes *can* be created manually, in which case someone reads through the document and creates the index by hand, typing every entry and page reference in a separate document. Before computers, this is how all indexes were created; some freelance indexers still want to do it that way. We advise against it because there's simply no reason to approach indexing that way anymore. Your desktop publishing tool or word processor is nearly guaranteed to come with indexing tools; typically, you place index markers in the text next to the terms you want to appear in the index and the software takes it from there. If your regular program doesn't include an automated indexing tool, buy one.

Dodging Bullets

You might have heard about or seen "joke" entries in indexes (such as the famous entries for "loop, endless—*see* endless loop" and "endless loop—*see* loop, endless"), but avoid creating such entries in your indexes. Your user is seriously searching for information, probably in a stressful situation. Like other forms of humor in technical documentation, it wears thin fast or could even misfire completely.

Pros Know

Not everything in a document should be indexed. The following items do not belong in your index:

➤ Information in the front matter; this includes the cover, copyright page, table of contents, or revision information

➤ Bibliography listings

➤ Irrelevant mentions of a topic

Another reason not to do indexing manually is that you won't be able to maintain it without calling in a indexer every time you update the document. You'll probably end up placing index markers into the text sooner or later anyway. Why not do it that way to start with?

Should Web documents be indexed, too? Yes. Create PDF documents with Acrobat data so the index markers are "activated" (they will jump the user to the reference point). If you use HTML documents, create an index with hypertext links. The concepts of indexing are the same for online and print documentation; the only difference is whether the reader turns physical pages or just clicks a link.

If you're not someone who finds it easy to index as you write, help yourself get a jump-start by sitting down with a hard copy of your document and a highlighter. Go through the document and mark the page every time there is a term you want to index. If the term suggests a concept that is not worded that way in the paragraph, make a margin note about what term to use in the index and that you need to point to this page.

Some Rules for Indexing

The following rules make it easier to create an index that is more likely to be right the first time. Requiring that your index entries be consistent lessens the number of "repair cycles" to tweak the final result after the software generates it.

➤ Use lowercase letters for every word except proper nouns.

➤ Use plurals wherever possible, unless the item in question is one of a kind, at least within your product. For example, use the topic "servers," as a plural, but write, "Web server" if there is only one Web server in use.

➤ Use *gerunds* (verbs with -ing endings), such as "installing" and "adding," for your verbs, to focus on the action.

After you have marked the hard copy, go through the document files and place the index markers. As you progress, expect to find more places to add markers than just the ones you highlighted.

When you think you're finished, generate the index and print a hard copy. You'll be surprised and probably dismayed to see how many errors there are in it. Don't worry, that's normal; the proofing and tweaking stage will take several iterations before your index is perfect. (See? We told you it takes a human to create a good index.)

Common Indexing Errors

Two of the most common and most regular errors in the initial index iterations are ...

➤ Capitalizing an entry in one place but not another ("System Administrator" and "system administrator"). The indexing software will not merge these entries;

capitalization is read as a different spelling by the indexing function. Consistency is crucial.

➤ Mismatching verb and noun forms ("configuring" and "configure" and "configuration"). Choose one form and stick with it (use "*see*" entries if needed).

Correct and generate the index as often as you need to—nobody's counting. When it's finally finished, it will be a great accomplishment. Do something nice for yourself—you deserve it!

Master Indexes

Master indexes are indexes that cross more than one document, often a whole collection of documents. These can be difficult to create but are enormously useful. If your product provides a multiple-manual set, often it is very confusing for users to figure out which item is in which document. For example, we once helped produce a 70-document set for one telecommunications product. The master index was one of those documents, and it was also re-created online; clicking on a hyperlink in the master index caused the referenced document to open to the place where the index marker had been created. Can you imagine trying to find your way among 70 different documents? Customers would have been completely at a loss without a master index, and indeed the value of the whole document set would have been substantially diminished.

Be aware that not all word processing or desktop publishing tools let you create a master index. If a master index is something your company needs, investigate the big names in desktop publishing software to find one that offers master indexing capabilities. Be careful to search for and ask about *master* indexing, not just indexing.

Also, be aware that even with good software, it takes a long time to create a master index and it is difficult at best. Think about why this must be so: All writers on all documents must use exactly the same terminology (including capitalization and spelling) for their entries. The index requires consensus and rigid consistency among writers, which means repeated, in-depth meetings. You'll also need to have a fairly specific style guide to indexing formats in place before you start (don't try to develop one as you go).

But whether you're building a three-page index to a product brochure or a master index across tens of documents, the effort is always worth the work. To users, a good index is worth *your* weight in gold.

The Least You Need to Know

➤ Include an index if your document is four pages or longer.

➤ Create a good index by putting yourself in the user's place and thinking of gaps each piece of information might fill.

➤ Take advantage of your word processor's or DTP's automated indexing tool, or buy one to use; don't create your index by hand.

➤ Create a master index for multiple-document sets—your users will be glad you did and lost if you didn't.

Chapter 15

Making the Final Laps

In This Chapter

➤ The joys of editing, rewriting, and proofreading

➤ Testing for quality assurance and usability

➤ When to say when

➤ Final delivery

You're making the final laps leading into the home stretch on your document. Just a few more times around the track for final editing, proofreading, and document testing, and you'll be ready to deliver the finished document. Kiss the trophy and smile for the cameras!

But the last few laps of a race always hold the element of surprise. Will everything go smoothly? Will something go wrong? If you've planned for the process and thought your way through to this point, you should be able to finish up with no problems. As you'll see in this chapter, this is where it all comes together—where the rubber meets the road.

Rewriting and Editing

You've finished writing and indexing and the document has been reviewed and approved by this time, so you know the content is solid. Now's the chance for a final pass (or two or three) through the document from a strictly editorial standpoint. We hope you'll be doing only some minor editing at this point, but there might be sections that need rewriting to be more polished.

You might not think this would be the case, but you actually need more focus and discipline when you're rewriting or editing than when you are writing in the first place. When composing your draft and making changes and additions for content, you have a certain freedom—it's a discovery process as much as a construction process, and you can rearrange things as you go. When editing, you're bringing a much more critical eye to the document, paying attention to those rough spots you skimmed over before. In rewriting, you must improve the section being rewritten without destroying the way how that part fits into the flow of the whole document. Editing and rewriting both must be done without damaging the coherence of the document.

Pros Know

"I've been involved in a lot of telecom issues addressed by industry committees and the FCC, and documents are always modified from the first version. Reviews are how we build consensus, and you have to see something in print to know what you've got. It takes many reviews to reach a final version, but when we do, we know it's solid."

—Judy Cook, Program Manager, Carrier Command Center, Sprint Corporation, www.sprint.com

How Much Is Enough?

It's a judgment call about how much rewriting or editing to do. If you were really rushed in putting the document together, you might look at it and feel as if *all* of it needs to be rewritten. Although there might be some truth in that assessment, you simply don't have time to rewrite everything for this deadline. You have to pick and choose.

As editor, you are the reader's advocate and stand-in. Focus on things that affect how clearly the meaning of a sentence, paragraph, or chapter comes through. Are any word choices ambiguous? Change just those words to something more explicit. Has any inappropriate passive voice crept in? Rewrite the sentence to eliminate it. Is there a single, clear idea in every paragraph? Split up any large paragraphs that have two, three, or more ideas in them. Do you have a clear lead-in for each set of instructions or procedures? Add a sentence or two if you don't.

Maybe you look at a sentence or paragraph and have a vague sense that it isn't as good as it should be—but you can't put your finger immediately on what needs to change. Leave it alone. If the problem doesn't stand out for you, it probably won't stand out for the reader either. Keep your attention on the things that do stand out.

Keep Things Moving

Balance your focus with good momentum as well. When editing or rewriting, it's easy to get bogged down in a particular section or even a single group of paragraphs. Keep your perspective on the document as a whole, and keep moving through it. If you find a part that cries out for more attention, place a bookmark or sticky note to keep that place and move on. Come back to it only after you've been through the whole document at least once. Your drive is for a consistent level of quality, not a mix of diamonds and rocks.

Then, go back to those difficult parts and address them in order, first according to their importance to the document and second according to how quickly and easily you can fix them. For example, there might be one troublesome section about a key process and a few short paragraphs that desperately need to be rewritten in an optional reference appendix. Focus on that key process first—it will have the greatest overall impact for the reader.

Editing Your Own Work

It's much harder to edit your own work than someone else's. Why? Because if it's your own work, you know what it is *supposed* to say and might not be able to tell whether it really succeeds. You're also used to seeing any gaps or flaws that might be there and you might not notice them anymore. Simply put, it is hard to be objective about your own work. Furthermore, you just might be tired of looking at it by this time!

Avoid editing your own work if you can. Writers in a department commonly perform *peer editing* for each other. Not only does the editor bring a fresh eye to the manuscript, but it helps that editor learn about the document and its product. This is important because document ownership might shift for the next release.

If you are stuck and must edit your own work, make yourself put it aside so you can gain some distance from it, even if it is just for a single day or even for a few hours.

Tech Talk

Peer editing refers to editing done by a colleague who is at an equal standing within a company or organization, rather than farther up the hierarchy. One advantage of peer editing is that it prevents an editing bottleneck within a department. Another advantage is that it keeps the focus on the document rather than introducing the element of relative corporate position into the editing process.

It also helps to break the editing process into several sessions if you can, because as you read it you become close to it again. If you turn three to five pages without wanting to make a mark on the page, pause and ask yourself if there really is nothing in those pages that could be improved (whether you intend to improve it or not). There's a difference between the choices made during "editing triage" and simply not seeing something that needs attention. It might be time for a break to regain your distance.

If you're already confident about the overall flow of the document, you might want to edit the parts or chapters out of sequence. This can be helpful in breaking out of that familiar groove when reading something you've already looked at too many times.

Editing Another Writer's Work

Although it's much easier to edit what someone else wrote, there are pitfalls in editing others work, too, that you need to avoid. Perhaps the biggest trap in editing someone else's work is being tripped up by differences in personal writing style. Although it's true that all of a department's documents should have a similar feel and follow the same style conventions, individual differences between writers inevitably creep in—especially if the document was just originated, or was heavily revised or completely rewritten. As an editor, keep your focus primarily on the content and on how the document complies with department style—not on whether this sentence or that paragraph is how you would have written it.

If you come across a passage and you're not sure exactly what it means, don't just take your best guess and revise it to match (unless you are absolutely out of time to do anything else). Unlike the reader, you have the writer available—ask for clarification, then edit or revise to reflect what you learn.

Sometimes you can quickly make vast improvements in someone else's writing simply by tightening it up. This can mean simply cutting out unneeded words, or recasting a long sentence into one or two that are both shorter and more direct. Look for wordy constructions such as the ones shown here that are quick to fix and make big improvements:

➤ For "in order to," write simply "to"

➤ Replace a construction such as "in the configuration of" with the more direct "in configuring"

➤ For most instances of the pseudo-scientific "utilize," write the straightforward "use"

You will find it useful to review Chapter 19, "Writing Clearly," right before you sit down to edit someone else's writing—and so will whoever edits your work. In the throes of creation and the pressure of deadlines, few of us write in a way that can't be improved by applying those guidelines afterward.

We also want to mention that a sense of consideration—we might even say compassion—also is important in editing someone else's work. As a professional, that person probably has little or no emotional attachment to the contents of the document. But who doesn't like to feel that a peer has respect for his or her work? Even though a SME's editorial comments might be accepted with a shrug, don't assume all yours will be. Treat your colleagues with the same respect and tact you'd give your most important SME and that you'd like to receive yourself.

Proofreading

Do you drive your friends crazy in restaurants, pointing out typos on the menu? Does it bother you when grocery store ads say "everyday" when they really mean "every day"? If you're grinning sheepishly and nodding, you probably won't have any problems with proofreading. If, on the other hand, you're one of those friends *being driven* crazy and you never notice anything about grocery store ads except the prices, proofreading is likely to be a challenge for you.

Proofreading means reading through a document on paper and paying meticulous attention to fine details—spelling (even if you use a spell-checker), punctuation, grammar, capitalization, correct use of fonts, correct titles and captions for illustrations, even the spacing between paragraphs and headings and after periods. Think of proofreading as going through a document with a fine-toothed comb, smoothing out the last little tangles and making everything tidy.

You might be wondering why we say to do this on paper rather than onscreen. After all, if you do it onscreen, can't you just make the changes as you find them? We insist that proofreading be done on a printed copy because we guarantee you will see things on a printed page that you will miss onscreen. Part of the explanation for this has to do with the higher resolution of print over text displayed on screen, but that's not the whole story.

There's something about having a concrete, tangible object in front of you that makes it easier to see things that need to be fixed. It's also easier to find your place again if—or should we say when—you get interrupted. Last, if you mark up a printed copy, you can hand it off to someone else to put in the actual changes while you go on to something else.

The following figure shows how to mark up text in a more or less universally accepted way so the meaning of each change is clear.

Pros Know

If you have to proofread or edit your own work, wait as long as you can between finishing the writing and starting the proof-reading. If you can, work on another project in the meantime. This distance allows you to see your work with a fresh eye.

Dodging Bullets

No matter how confident you feel, make your proofreader's marks in pencil, not pen. Then, if you change your mind, it's easy to change your marks, too.

Mark	Meaning	Mark	Meaning
ℒ	Delete	ⓢⓟ	Spell out
ℬ	Close up (delete space)	⫽	Use lowercase letters(not capitals)
stet	Let it stand	b̲f̲	Make bold
#	Insert space	⩣	Use capital letters
¶	Start new paragraph	⋀	Insert comma
]	Move to right	⋁	Insert apostrophe(or single quote)
[	Move to left	⊙	Insert period
][	Center	M̅	Insert em dash
tr	Transpose	wf	Wrong font

Standard proofreader's marks.

By the time you get to proofreading, you're not thinking about content anymore. For example, you're not asking whether an illustration is current or correct, only whether it is correctly captioned and whether the caption is correctly capitalized and punctuated.

Proofreading also means checking every cross-reference to be sure it points the reader to the right place. You might feel this is redundant if your software manages cross-references for you automatically, but it's worth doing. A cross-reference marker might have been accidentally moved when other text or graphics were cut and pasted.

Coffee Break

STET (pronounced *stet*) is a common proofreader's mark that confuses nearly everybody the first time they see it. It is used to reinstate text marked for change or deletion and means "let it stand." You might expect it to be an abbreviation or an acronym, but it is, in fact, the present subjunctive third person singular form of the Latin verb *stare*, which means "to stand." You don't need to know this to use it, of course, but someday someone is bound to ask you what it means—and now you can give a good answer.

You'll also check the table of contents to confirm all page numbers are accurate. Usually your word processing software or DTP system has automated features to generate your table of contents, but it still needs to be checked by a human being. After all, a human being will be using the finished document.

The index should have been proofed and checked while you created it. (For more about creating your index, see Chapter 14, "Indexing.") Take another look at it, though. Make sure it isn't missing any start or end pages in a range of pages and doesn't contain the same entry twice (this happens if you make two index entries different when they should be exactly the same). Scan it to make sure it isn't missing anything important.

Making Proofreading Easier

If proofreading is something you struggle with, there are strategies you can use to get yourself through it. One of them is, believe it or not, to read the document backward, either word by word or paragraph by paragraph. Start at the last page and work your way back. This puts words out of sequence; undistracted by their meaning, you are able to focus on proofreading.

Another tactic is to put a white piece of paper across the page, positioned so the edge is under the first line of text. Then, simply move the paper slowly down the page as you proofread. This slows you down and makes your reading of each line more deliberate. It also isolates each line, which makes it easier to give it the attention it needs.

It also helps to print the document so the type appears a little larger—if you can do this in an easy, controlled way. For example, if your text is 11pt, bump it up to 14pt or 16pt. Be careful with this, however—make a copy of the document files and enlarge the copies, not the original files. Changing type size will wreck the pagination. It might be easy to change it back, but why take the chance?

Testing the Document

Often this step is the one that's cut if the deadline pressure is fierce, but that doesn't mean there's no value lost in omitting it. On the contrary, without document testing, the document's usability, clarity, and usefulness to the reader are based entirely on speculation. Informed and educated speculation, no doubt, but speculation nonetheless. Until a real user sits down with the document and uses it successfully, there is no certainty it will do what it's meant to do.

If you are fortunate, your company's usability or QA department incorporates document testing (including online help) into the release schedule, usually as the final step. During this phase, the tester goes through all user documentation and steps through procedures, checks to see that the screen captures match the software, and if you are lucky, reads for clarity and correctness as well. Unfortunately, not every company incorporates this phase into the process. Tight deadlines in the high-tech world and the short supply of qualified QA staff make this a rare luxury indeed.

Document Testing on a (Time) Budget

There's no way around it; those improvements will take time, and time is what the high-tech industry has the least of. Is that justification for skipping document testing? No—it's justification for being sure to include it in your planning and scheduling from the outset.

If you have to do your own document testing, enlist the help of your colleagues in your tech writing group. If there's little time to spare, try to incorporate a testing phase wherever it can fit in, rather than waiting until the end.

A Testing Checklist

Make sure the document has been found to be acceptable on all of the following points:

➤ The user is able to find information quickly.

➤ Instructions are clear and easily followed.

➤ The instructions work correctly.

➤ The user can find his or her place quickly on the page after glancing away.

➤ The user's questions are answered within the document.

➤ The index is thorough and contains sensible entries.

A Final Document Checklist

As a last stage before you turn the document over to production, you should go through a checklist of details. The final checklist is to ensure that the little things are done and the final product is as good as it can be. Here are some items to include on your checklist:

➤ Run your word processor's or DTP's spelling checker.

➤ Remove revision bars and comments.

➤ Check all page numbers, headers, footers, and chapter numbers.

➤ Make sure you followed the style guide.

➤ Make sure pages break correctly.

For online documentation, your checklist can include items such as …

➤ Checking all links to see that they go to the right target

➤ Checking the index for completeness

➤ Making sure all online help is up to date

Pros Know

If your department doesn't have a final checklist, make one for yourself and share it with your colleagues. They will find things to add to it.

Freezing the Document

You might think that you would have stopped making content changes to the document long ago, after the final review sign-offs. Theoretically, that's true, but pragmatically things just don't work that way.

A SME might tell you that something he or she told you earlier has now changed; the product might undergo last-minute changes that mean what you've written must be adapted to match. These and other events like them come with the territory of technical writing.

But you reach a point where you have to shut the doors on changes if you're going to finish the document and have it ready in time to meet your deadline. This is called "freezing the document."

Collecting Post-Freeze Changes

Do the requests for changes to the document stop coming in? Of course not. But if the document has been frozen, those changes must wait for the next document version or the next product release to be incorporated. Be sure to collect and organize them so they'll be available to whoever has ownership of the document for the next round of revisions.

It's essential to make sure SMEs and others know when the document will be frozen. They also must understand that when you say the document is frozen, you really mean it. We're talking frozen like a block of ice, not like a pile of slush.

Don't Thaw—Add On

There might, however, be truly a crisis that calls for something to be corrected or added after document production is too far advanced for you to fix it and still meet your deadline. In that case, remember you can include an addendum as a separate sheet in printed manuals, and in a README file for documents viewed online or from a CD, to provide that last-second information. Release notes also often contain documentation errata. Remember, your document is frozen.

Handing Off the Final Deliverables

The document is finally complete and you're galloping across the finish line—sometimes literally, as you race around the office from computer keyboard to laser printer to photocopy machine.

The next step is to hand off the document. This, of course, depends upon how your company does things. If you don't know, ask! You don't want to print out camera-ready copies on a laser printer and then find out you need to send a PDF file to a printing company's DocuTech system instead.

You probably will be responsible for handing off a completely finished, camera-ready document to your boss or a project manager. If you have to handle printing yourself, you will work out the format with your contact at the printing company.

You also might provide PDF or HTML file copies to someone who will put these copies on the company's Internet, intranet, or extranet site, or on a CD that accompanies software.

Whatever the final deliverable, make sure that you have checked everything. If you're the last one to have it before it goes to production, you're the one they're going to come to if there's a problem.

It's usually a good idea to keep all the paperwork for a document at least until the next release is completed, and sometimes even longer. Although it doesn't happen often, once in a while there will be a problem at a later date, and you might need the paper trail to prove you did what you were supposed to do.

Can you believe it? You've finally completed the entire product lifecycle for a document. You took it from infancy through the draft stage to final production. Congratulations! You deserve it.

Pros Know

Keep all the paperwork you've collected with a document: the document plan, your checklist, your review copies, and the sign-off sheet. Store them in a folder with the final printout of your document.

The Least You Need to Know

➤ Get fresh eyes to edit your writing.

➤ Be positive and supportive in editing and revising the work of others.

➤ Test your documents to confirm usability.

➤ "Freeze" documents at the right time—and mean it.

➤ Keep your paperwork until the next release is over. Although we hope you never need it, you'll be glad you have it if there's a problem.

➤ Pat yourself on the back for a job well done!

The Deliverables

Up until this point, you've stepped through the whole writing process without ever having a peek at an example of a "typical" document. We're not keeping it a secret, it's just that there's no recipe that tells you exactly what the ingredients are and what you need to write. Instead, you have to go through the processes outlined in this book—learning the product, the audience, and going through reviews—and as you progress, you gradually realize what information needs to be included. Does that make it impossible to understand what belongs in any given type of guide? No—there are certain characteristics you can count on to be present in various types of technical documentation.

To get ideas for your documentation, gather guides of various types for different products and study them side by side. You'll begin to notice patterns. Use the other guides to determine what your own table of contents should be—if it works for them, it can work for you, too, at least until you learn more about what your end users need.

You'll notice, too, how many of these documents overlap in their purpose. There are many reasons for that: Some companies don't want to pay for or don't want to give the customer too many different documents. Sometimes products are not complex enough to warrant many different types of documents. And sometimes there's just no time to write all the different types of books a product might need. Although we've attempted to categorize the documents and explain what type of user needs them, the fact is, any document might be used by any array of users.

As you develop experience, you'll write many different technical documents, and these categories will become second nature. To get you started, this chapter contains descriptions of some of the standard documents, so you know what to expect until your experience catches up.

User Documentation: The User's Main Course

User guides and user handbooks are where you serve up solid content, and plenty of it. We classify tutorials and "getting started" guides into the user guide category, because they, too, are designed for the end user.

These documents are the user's companions and references for working with the product on a day-to-day basis. They cover everything the user needs to know about every feature, every option, every customizable setting. Depending on your approach, they might even describe every screen—and every field on each. (See Chapter 5, "What Makes a Good Document?")

Coffee Break

Do you know the difference between a note, warning, and caution?

A **note** is a piece of information that is not crucial to the documentation, though it may be interesting. A user can skip this without losing anything essential. A **warning** means there is a possible danger to a person. A **caution** means there is a possible danger to equipment. It is also a good idea to explain what these icons mean in the introduction to your manual.

Many tech writers enjoy writing user's guides because they can easily put themselves in the end user's position. As we've said before, the tech writer is the first end user.

Standard User's Guides

A user's guide can include anything and everything, including installation information. If your organization delivers only a small set of documents with the product, it's very likely that a user's guide of some sort will be one of the documents.

In general, the purpose of a user's guide is to explain tasks that are done regularly. Write your user guide with this structure in mind: simpler things first, building toward more complex things later. A typical user guide has the following elements, presented in this "table of contents" order:

➤ **Introduction.** An introduction to the guide, which tells what the product is about and what information is covered in the document.

➤ **About this product.** Information about the features and functions of the product. You might not list every single menu item and dialog box (that can be put in a reference guide), but you must explain how the features are used.

➤ **Using this product.** Basic or fundamental content relevant to all users, such as tasks that are done most frequently, or that are part of other, more advanced tasks.

➤ **Features and functions of this product.** Advanced content about options or features that are the province of the power user who pushes the product to its limits.

➤ **Appendixes.** Reference appendixes that address topics not of interest to all users or that relate to add-ons or enhancements that not all users will have chosen. For example, you might include as appendixes keyboard commands, information about how to customize the product, service and support information, or extra applications that not all users buy.

"Getting Started" Guides

The purpose of a "getting started" guide is to get the user up and running as quickly as possible. A "getting started guide might stand alone, or might easily be part of a user's guide, an installation guide, or even a tutorial. It might be a quick reference card or it might be the only user's guide provided for a product. A "getting started" guide generally has the following attributes:

➤ Makes it clear what product the guide is for, but doesn't describe the product in any detail.

➤ Is brief, usually fewer than 50 pages and sometimes as short as a single page.

➤ Can consist largely of numbered steps.

➤ Helps the user start interacting with the product without having to learn much about it. For example, you might perform a few basic tasks, learning only enough about the windows and menus to help you do that task.

When planning, outlining, and writing a "getting started" guide, keep the document's purpose firmly in mind. Get a clear vision of the circumstances the user will be facing: Will the product already be installed, or does he or she need to install it? What list of actions needs to be completed before the product is ready to use or ready to be worked with in other ways? What information does the user need to have on hand to get started, and where does it come from?

You'll find yourself tempted to put in lots more information than people really need at this point. Capture these bits and pieces so they can be put where they belong.

Tutorials

These are like the start of a good meal—more than an appetizer, but not yet the main course. People sometimes think of tutorials and "getting started" guides as the same thing. And they might be, but they're not necessarily. As the name implies, tutorials are about teaching. Although getting started is something people seldom do more than once, users might come back to a tutorial again and again until they've mastered the features or processes it teaches.

A tutorial might stand alone, or might be part of any other user's guide. Tutorials come in many forms. At one end of the spectrum are simple paper documents with screen shots, created with tools such as Microsoft Word or PowerPoint. At the other end are interactive multimedia presentations with sound effects and built-in feedback and performance scoring, created with tools such as Macromedia's Authorware.

The form you choose is dictated by the balance you reach among three factors: what the user needs to learn, the budget and other resources you have to work with, and how much time you have to create it. (Remember to allow time in the schedule for testing and adjusting your tutorial.)

Because the tutorial is a learning experience, start with simple, fundamental tasks and build gradually in complexity. Reinforce just-learned skills by having them appear again in later parts of the tutorial.

Quick Reference Cards

These are the snacks on the tech writing menu. Quick reference cards provide concentrated morsels of information. These "job aids" serve as reminders of things the user has already learned or as a handy list of commands or processes.

A quick reference card is a single piece—no bound books, so no pages can fall out. Make quick reference cards a handy size; small is good, but not too small. It can be folded into three or four panels, and can even be as big as a poster for the users to keep on their walls. Text and diagrams need to be large enough to be readable at a glance. The physical material you choose also needs to match the expected use this card will get—heavy card stock is a minimum requirement. Laminated or plasticized cards are even better because they resist wear and are coffee- and soft-drink proof.

Although short, a quick reference card is not so easy to write. To be in the right frame of mind to plan and write quick reference cards, you might need to turn your tech writer hat around backward. Rather than full explanations, think "highlights." Rather than sentences and paragraphs, think phrases, words, bullet points. Rather than "white space," think "saving space." Simple diagrams? Yes. A reference matrix? Absolutely.

Good candidates for quick reference cards are things that are specific and explicit, but that do not need a supporting context to be meaningful. Graphics are good when they apply. Adobe Photoshop shows a graphic of the toolbar with all its icons identified, plus pictures of each palette. Octel Aria's quick reference card includes a full-color matrix of functions of all keys. And the quick reference card in this book contains a graphic of the "five steps to creating documentation."

Pros Know

A rough rule of thumb is that if the installation process requires 10 or more steps, give it its own guide or its own chapter in another document. This is especially true if any of the steps needs to be split into substeps.

Introductions

A product introduction is a starting point for a large document set. It gives the user an opportunity to learn about the product and the documents that accompany it. An introduction can be part of a larger user's guide, or it can stand alone.

An introduction might contain the following:

➤ Detailed information about the product

➤ Descriptive information about the purpose the product fills; how the user will deploy the product

➤ Product *architecture*

➤ Hardware/software requirements

➤ The file structure of the product

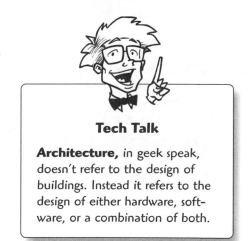

Tech Talk

Architecture, in geek speak, doesn't refer to the design of buildings. Instead it refers to the design of either hardware, software, or a combination of both.

Feel free to put more descriptive information into an introduction than into other user's guides. This is the document people read to learn about the product and the environment it is designed to work in. An introduction also might contain some user tasks, but only after the descriptive information.

Tech Talk

A **wizard** is a utility within an application that steps you through a particular task. For example, an "installation wizard" would take you through the steps of installing the product.

Tech Talk

Platform usually refers to a system's underlying hardware or software, but often is used to refer to the operating system.

Installation Manuals

Installation manuals are for everyone. Depending upon the product, you might be writing an installation guide for a naïve user or for a member of your company's professional services organization. Installation can involve anything from popping a CD into a slot and stepping through a few *wizards* to configuring numerous servers and installing several different software packages in specific order.

For simpler installation procedures, you can put this information in a "getting started" guide or even a different type of user's guide. For larger documentation sets and more complicated software or hardware, installation is addressed in its own document, an installation guide.

When writing an installation guide, make sure it includes the following:

➤ The configuration and setup required before starting installation procedures.

➤ Step-by-step procedures on how to install the product. (Avoid referring the user to other documents or Web sites.)

➤ An explanation of how to do the installation procedure for every *platform* the product supports. If there are procedural differences for UNIX, NT, Windows, and Linux, explain them all.

➤ Telephone numbers and e-mail addresses for customer and tech support.

Release Notes

Release notes accompany products when, as the name implies, they are released. Release notes provide the user with information that does not belong in or did not make it into the regular customer documentation.

For example, if you want to explain to the customer why this release is different from the previous one, you explain it in a release note. That wouldn't be appropriate for the main documents, because if a document is basically accurate, the same one can be used through more than one release.

Release notes also contain information about bugs or workarounds. This material also does not belong in the main document because the programmers hope to correct it before the document expires.

Release notes often are delivered as README files on a software CD. README files are typically produced as *ASCII* text for easy readability on all platforms; however, with today's common use of word processing programs or the availability of Acrobat Reader, you can produce these in PDF or another format.

Whatever format you produce release notes in, make them fairly short. A customer wants to read them quickly to see what's important and what's wrong. If there's any more detailed information, it can go into the regular manuals.

Release notes can include just about any type of information that the development team thinks should go in. You might include the following items:

➤ New features in this release and what they do.

➤ New documentation for this release.

➤ Bugs that haven't been fixed yet. (Don't call them bugs, though. You'll have to give them a more positive spin.)

➤ Upgrade procedures.

➤ Tech support information.

➤ Explanation of which files and directories will change after the new version is installed.

Troubleshooting Manuals

Troubleshooting material is for everyone: programmer, end user, installer, and system administrator. A troubleshooting manual contains descriptions of problems the user encounters, with information about how to solve those problems.

Troubleshooting guides often are difficult manuals to write as they require a lot of research. The writer

Tech Talk

ASCII (American Standard Code for Information Interchange) is a code for representing alphanumeric characters as numbers, with each letter assigned a number from 0 to 127. Because ASCII is used by most computers and printers, text files can be transferred easily between computers. ASCII is completely unformatted text, such as what you produce if you use Notepad and save it as a .txt file.

Dodging Bullets

There's really no good way to pretend a bug is a feature in release notes. However, there's no need to accentuate the negative. You can call the section dealing with this "Workarounds," or you can make a statement, with permission of the development team and product manager, that there are some "issues" that have not been resolved for this release but will be corrected in the next one.

needs to be a bit of a psychic here, guessing what troubles occur and then getting the solution. To find out what your user's problems are, consult a variety of sources. Your company's help desk is an excellent source of information. Every customer question should be considered for inclusion in a troubleshooting guide.

Troubleshooting guides, perhaps more than any other type of documentation, require your putting yourself in the user's shoes. Use the product yourself and note the stumbling blocks and error messages you receive. The programmers in your organization should be able to print out error codes and messages for you. Unfortunately, error messages are not always as clear as they should be. You might need some help translating them.

The Heavyweights

As a tech writer, you won't always get to write documentation aimed at an end user who is someone like yourself. You should also be able to write for more advanced users (what we called "power users" in Chapter 9, "It's All About Audience")—for programmers, installers, network administrators, and other "techie" types. These documents might include APIs, reference materials, and specifications.

These are the more technical types of documentation you'll encounter in your tech writing career—and the documents you have to be careful about. Unless you yourself have programming experience, you cannot bring the same end user experience to the documentation, as you can with user's guides. You must write solid, accurate reference information without a lot of user-friendly "fluff."

Application Programming Interfaces (APIs)

Application Programming Interfaces (APIs) are collections of routines, protocols, and tools that a program uses to direct the procedures performed by a computer's operating system. For example, the Windows API enables a program to perform actions such as opening windows or files, defining a mouse click, and browsing. Programmers use APIs to adapt a product to their own programs and systems.

The structure and content of an API document depend on the components in the actual API itself. For example, your organization might produce software that can work with another application such as e-mail or a word processing program. Another company will then use the API to learn how to customize your software to work with their own e-mail. All programs end up with a common interface.

The API document must provide all the information necessary to enable another application to interface to it and to perform all important tasks associated with the API. The document should list all functions and their meanings.

The best way to successfully tackle an API document is to find out what the programmers in your company believe should go into it. These programmers are identical to the document's end users! All the better for you if you are able to read the code; otherwise, you are simply formatting what the programmer provides for you.

Reference Guides

Reference guides are informational, nonprocedural documents. Reference guides are never task-oriented: They are used like a dictionary or encyclopedia, and referred to when the reader needs information about a particular topic. A reference guide might be a list of commands, error codes, or descriptions of all the GUI elements and what they do.

Producing a good reference manual requires mostly that you compile all existing information pertaining to the topic of the document. For example, in a command reference, you need to list every command, its meaning, and all options that apply to it. In an error code manual, list all error codes, their meanings, and possibly their causes and appropriate responses to them.

Note that a reference guide, depending upon its topic, might be used by more than programmers—we include it in this section only because it contains no user-oriented procedures. However, a glossary is certainly a reference and can be used by any kind of user.

Specifications

Specifications are plans that define what a product will do and, often, what it will look like. Specifications usually (but not always) are written before the product begins to exist. Once all the features and functions are agreed upon according to the specification, the developers then develop the product to match what was defined. Software developers often write specifications ("specs"), although in some organizations, the tech writers are responsible for this task.

Many small companies, startups, and Internet companies do not write specifications. Larger, well-established organizations almost always do. If you are not responsible for writing specifications, they will still be a large part of your life, because you'll use them as source material for your product documentation.

Your company might produce different types of specifications under different names. The following are typical:

➤ Functional specifications (or product specifications) describe what a product should do and how it should do it.

➤ User interface specifications define the user interface. These specs should show examples of all the parts of the user interface and describe what they do.

Pros Know

"When I started writing documents for system administrators, I
learned that while users prefer
easy-to-use products, system administrators prefer things that are
complicated, cool, state-of-the-
art, and hard to learn. Difficult
things present a welcome challenge for system administrators,
so if they have to resort to documentation, they want it to be
detailed, technical, and too difficult for regular users."

—Anne Karppinen,
Documentation Product
Manager, Nokia Networks,
Operations Support Systems,
www.nokia.com

Coffee Break

According to the Bureau of Labor
Statistics, the fastest-growing employment sectors over the decade
between 2000–2010 will be
high-tech and e-commerce.

If the specifications are written to conform to a type
of standard (see the description of ISO 9000 documentation later in this chapter), the content is clearly defined by the organization that governs the standards
you are following.

System Administrator Guides

System or network administrators are responsible for
day-to-day maintenance of the computer operations at
a company. They need task lists, relevant command
information, and maintenance information. The documentation might include both software and hardware
information.

The information in a system administrator document
depends on the operating system or systems in use. At
a minimum, a system administrator document should
contain information about the following:

➤ Regular operations and maintenance

➤ Hardware management

➤ User administration

➤ Networking

"Oh, and Could You Write This, Too ...?"

A tech writer's life isn't all software installation and
network configuration. As one of your company's resident writers—or maybe the only resident writer—you
might be called on to do anything that requires better-
than-average language skills. You could be asked to
edit a non-native English speaker's specifications, write
a speech for the company president, script a video
training session, or proofread your boss's annual presentation. Here are some of the many projects you
might be asked to contribute to or write:

➤ **Marketing communications (marcomm):** This could include writing press releases, revamping product spec sheets, or editing the annual report. Marcomm, as it's known, calls on many of the same skills tech writers use in their usual documents, but requires writers to use snappier, more colorful language.

➤ **White papers:** Often a white paper mixes marketing and technology to explain how a product works and why your company's solution to an industry need or problem is better than that of their competitors.

➤ **Training materials:** Like user guides and programmer documents, training materials need to be scrupulously accurate about all aspects of the information they provide. These are materials that teach people, and the user depends on them to convey the information correctly.

➤ **Professional or reference books:** Some of the software companies now don't include documentation with their products, but rather depend on retail books to explain their use. If you know some of these product very well, there's no reason why you can't be writing a book for a national publisher rather than just for your company.

➤ **ISO 9000 documentation:** ISO 9000 is the shorthand name for a group of standards governing manufacturing processes. When a manufacturer complies with the standards, it becomes ISO 9000 certified. One part of the ISO 9000 standards mandates there must be a quality assurance program in place with documented polices, procedures, and work instructions. If you're a tech writer at a company producing ISO 9000–certified products, it might fall to you to write the documentation.

As you saw in this chapter, you could be expected to produce many different types of documents for a diverse array of users. It's a good idea to read widely and become familiar with all these different document types; sooner or later, every tech writer's middle name becomes "versatility." But every deliverable, no matter how far off *your* beaten path, must still incorporate the fundamentals: It must be complete, accurate, readable, and focused for the intended audience.

The Least You Need to Know

➤ Study examples of each document type, and borrow structural elements from them to get you started.

➤ Organize information based on how people learn—start with simple ideas and build toward complex ones.

➤ Read and become familiar with a wide range of documents; tech writers are sometimes called upon to show their versatility, from writing marketing copy to authoring trade reference books.

Part 4

Knowledge Is Power

Truer words were never spoken—or written. One key to your success as a technical writer is the level of mastery you can achieve with an array of software tools.

A word processor or desktop publishing system might be your daily tool of choice, but you'll also have opportunities to use graphics software, online help generators, HTML page tools, and others, all on the same document or set of documents.

Adding breadth and depth to your knowledge of tools is power in another way, too—it gives you a wider field of options for the kind of work you want to do and where you can steer your career.

"You Want It When?"

In the computer industry, everything has to be done "yesterday" if not "last week," and it seems there's never enough time to do a job right. But doing a job right depends on knowing what "right" looks like.

With effective planning you can produce a document that is pretty good, fast, and inexpensive under virtually any time constraints. In this chapter we tackle the toughies: how to gauge the levels of effort in a document, how to calculate the hours it will take to write it, and how to plan your time around a deadline. Remember, the clock is ticking!

"You Can Have It Good, Fast, or Cheap: Pick Two"

You've probably heard this saying or seen it on a sign in a shop: "You can have it good, fast, or cheap: Pick two." That saying has been around for a long time, and like a lot of things that have been around for a while, there's a lot of truth behind it.

But times do have a way of changing. In the computer industry, companies want their documentation to be good … and fast … and cheap. But this flies in the face of conventional wisdom—is it really possible?

Yes, it is possible, but only if you amend the saying a bit: "You can have it *better*, *faster*, or *cheaper:* Pick two." See the difference? In this saying, things aren't either/or—instead, the three characteristics are relative. It remains perfectly true that some aspects of each one will have to be sacrificed, but the result will never be bad, slow, or overpriced.

Here are some examples of how this dynamic relationship plays out: You can often provide speedy, high-quality work—by paying for more than one writer on the project or paying for overtime. You can supply an extremely high-quality result by using a single writer—but that writer needs a reasonable and uncompressible amount of time to do the job. You can do a job on a crazy schedule and do it all by yourself—but the resulting document won't be the one you know you're really capable of writing.

What Do They Really Mean?

Instead of saying that doing a job well, quickly, and cheaply is impossible (we know that's a word your boss doesn't ever want to hear) consider what those three words mean in the technical documentation world. Those three words are strong factors in determining the scope of the project, but you need to figure out which of these criteria are the most important.

Writing a document well is always important, isn't it? Or is it? If you ask a documentation manager how important it is to produce a high-quality piece, we'll bet he or she will say it's always important. But events don't always bear that out.

Pros Know

Experienced writers know that the finishing touches—rewriting, proofreading, editing, indexing, preparing for the printer—always take far longer than expected. Allow plenty of time for "cleanup" work or have someone on tap to help you with it (it's easier to delegate that part of a project).

To gauge how important quality is, observe a situation in which the documentation team, for whatever reason, can't complete the documentation or online help in time to meet a tight deadline. Does the manager push back the deadline and tell the writers to continue working until the document's content is complete? Or are the writers instead instructed to put the document out in whatever form it's in, and wait until the next release or revision to add whatever is missing? If it's the first, quality is more important than speed; if it's the second, speed is the higher priority.

But is speed the *most* important consideration? In an industry in which time feels measured in microseconds—or in nanoseconds, at Internet-driven "dot-com" companies—speed almost always is the most important of the three criteria. To confirm this at your company, determine how the documentation is being delivered to the customer. Will it accompany the software on a CD?

And when is the target delivery date for the software? If it's soon, and usually it is, there can be no doubt that speed is crucial—your documentation *must* be completed simultaneously with the software, even software that changes up until the last minute. At some companies, they'll hold the software release until accompanying documentation is finished, but this seems to be the exception rather than the rule.

Cheap isn't necessarily as bad as it sounds. It refers to the amount of money—often, but not always, measured primarily in hours of writer pay—required to get the job done well enough and fast enough. How do you know where this characteristic falls on your company's list of priorities? Ask yourself this: If a pilot installation of your company's product were happening on the other side of the country, or of the planet, and the documentation would significantly benefit by having you observe the process, would they send you? If the answer is "no," you can bet your management is more interested in cutting costs than improving quality and perhaps speed.

Or, more realistically, suppose a project is due in an impossibly short time and your only hope is to hire two experienced contractors to help you. If your management agrees to hire them and doesn't quibble about their rates, you know speed is more important than price.

On the other hand, does your company agree to the cost of adding functional features such as color inserts and tabs to a printed document? This tells you that quality is uppermost in their minds.

Battling the Inner Perfectionist

When speed and price are more important than quality, and they often are, expect to be frustrated. This is because it seems to be part of the tech writer's nature to want things to be perfect. And that's not surprising; after all, we're detail-oriented, we like explaining things to people, and we're interested in accuracy. We care about the placement of commas, consistency of the edit, visuals, and other details. It's part of what makes us good at our jobs.

But if, as commonly happens, speed and cost are more important than quality, you'll have to learn to let go of that vision of perfection—without giving up on quality completely. To see how to do this, think about quality as consisting of five factors, listed here in order of importance:

1. Accuracy of content
2. Completeness of content
3. Consistency in style, typography, and correct spelling
4. Layout as it affects usability and accessibility
5. Presentation enhancements related to paper thickness, use of color, graphic elements, and packaging

Much as it pains us to say so, we've learned through experience that you can in fact do without 3, 4, and 5 in a technical document. But you absolutely cannot do without 1, and having 1 might not do you (or the user) much good without 2. That's the bottom line.

When Speed Is of the Essence

It will always be true that some people work faster than others. Even though as a slow starter or methodical worker you might never set land speed records for your technical writing, there are tricks anyone can use to work faster:

Know thy subject. We've been through this before, and it will come up once or twice more in this book: If you're well-versed in the product and how it works, you'll find it a lot easier to have something to say about it.

Do a little bit each day. We don't know why, but we know it's true that every writing project takes longer than you think it will. And if, by some miracle, it doesn't, think how nice it will feel to finish ahead of schedule! Increase your chances of feeling that smug satisfaction by starting earlier than you think you need to, and by working on the project regularly even if you don't want to.

Inch by inch, writing's a cinch. Break down your schedule into smaller milestones than the major ones looming, and base them on a real schedule. For example, if your document is due in eight weeks and you have 12 chapters to finish, mark off milestones within those eight weeks, then divide even those milestones into smaller ones. You might plan to finish two chapters a week and then spend two weeks on reviews, indexing, and final edits. Make the tasks even smaller by assigning yourself one chapter every two and a half days, then a specified number of pages per day. This also helps you work at an even pace—"steady" really does win the race.

Create an emergency. We're not kidding. If you're one of those people who waited until three days before a college term paper was due to start writing it and still got A's, maybe you need to create a time crunch for yourself. Not surprisingly, this is quite easy to do: Send out invitations to a review when you haven't yet completed the document, and tell the reviewers you'll be sending a draft copy on a certain date.

Multitask. Remember the circus sideshow where a performer sets plates spinning atop slender sticks and then rushes madly from stick to stick to keep them all spinning? This is the essence of multitasking, and might be the true secret of people who work at lightning speed.

Pros Know

You don't always have to go by your own personal experience to estimate timelines. Here are useful parameters from industry studies for how much time to budget for creating online help: three to six hours per screen when extracted from a manual; four to seven hours per screen if done from scratch.

When a "unitasker" working on a document gets blocked or tired and can't seem to move forward, he or she might stare at the screen, visit the vending machine, or go to chat with a co-worker for a while, waiting for the block to shift. Eventually it does, and the writer comes back and works some more.

A true multitasker might experience these same episodes of fatigue, but responds to them with a different strategy. He or she switches gears and finds refreshment in changing one task for another. Instead of doing something unproductive, the multitasker who needs a break jumps into a task that is significantly different from the original task, such as indexing, proofreading, or writing a different document.

Dodging Bullets

Be careful! There's an important difference between creating motivation for yourself and setting yourself up for failure. If you don't come through on what you've committed to, colleagues will think of you as someone who doesn't follow through, and reviewers won't feel there's any particular rush to respond on time when they *do* get the documents.

Working at breakneck speed means doing something productive all the time, and often two things at once: These people read while waiting for the system to boot up, type e-mail while a file is being saved, edit and proofread while talking on the phone. In the same time it takes to eat a bag of chips or hear about a co-worker's vacation, the multitasker has progressed on a different task, and might even have finished several.

Luckily, we don't all have to be like that. But you may be able to help yourself work more efficiently by trying some of those techniques. If you can dovetail several different tasks together, every minute is used productively.

Dodging Bullets

If you're not naturally the type who fills every moment with two tasks going on simultaneously, an intention to multitask can wickedly transform itself into diversionary procrastination. You can fall prey to that urge to do something—anything—other than what most needs to be done at the moment, and instead of switching to an important task, might find yourself surfing the Web or reading a trade magazine. Set time limits for your breaks and make sure you return to the main task in no more than 30 minutes. (Use a timer if you have problems with procrastination.)

Working with Contractors

When tech writers need help and their budgets allow, it's a common practice to hire *contractors* to help out. As the saying goes, many hands make light work. (By the way, if you're planning to become a contractor, this section is important to you, too. It always helps to understand things from the employers' point of view. Chapter 23, "Office Alternatives: Working Outside the Box," discusses more about contracting for those interested in getting into it.)

Tech Talk

Contractor refers to a professional who works on a contract basis, often as a subcontractor for a general project contractor and usually for a specific project and with a budgeted number of hours. Expect a contractor to be able to come in and do a job with little or no training time. Contractors typically are paid a straight rate for the hours they work and receive no benefits (vacation, paid sick days, insurance, and so forth) from the company buying the contractor's time.

But this also can be an unexpectedly expensive way to get the job done, and more time-consuming than time-efficient. Although a good contractor should be able to step in and hit the ground running, there's no way around the time it takes for someone new to come up to speed. Hiring two extra people is not like cloning yourself twice—sometimes it takes two contractors to equal one of you. You might not realize how much you know—and how much contractors don't.

If you're under pressure, the deadline is close, and you're thinking of hiring contractors, step back and take a breath. Are contractors really your best bet for meeting the looming deadline? Can you afford to spend a considerable percentage of your remaining time acquainting them with the product, explaining the documentation, and making sure they understand what needs to be done, and how? Once the contractors are productive, will there still be enough time to get the work done? The answers aren't always "no," of course, but they aren't automatically "yes" as often as you might think.

The way out of this sticky situation is not to get into it in the first place. This means keeping closer tabs on how a project is progressing and anticipating your needs so you can ask for help before the moment you need it. If contractors need a week to come up to speed, you can spare that time when no deadline is looming, but not when the deadline is two or three weeks away. You'll get many weeks of productive work, instead of just one or two, by "investing" that first week sooner rather than later.

Coffee Break

For those of you who think you can save time by throwing more people on a job: You can't have a baby in three months by putting three women on the project.

If budget allows, hire a contractor when things are quiet and put him or her to work on something necessary but not time critical. This helps you and the contractor learn to work together and acquaints him or her with your company, the product, and the documents involved. Then just hope that person is available the next time you need extra help!

If the project schedule changes because of things outside your control, however, you might have to bite the bullet, hire some contractors, and make the best of things. But less is more when it comes to extra people—remember, the distractions for you multiply by the number of contractors you add. The key is to find a balance.

When you need to hire contractors, bear these points in mind:

➤ Make sure the person you hire is familiar with the tools you use. It's not cost-efficient for you to pay someone to learn UNIX or FrameMaker.

➤ If it's a short-term project and the deadline is tight, don't expect contractors to learn much about the product. They simply won't have time. You'll have to figure out what parts of the project they can work on—such as editing, rewriting, indexing, stringing files together, or creating online help from existing documentation—while you keep writing.

➤ A highly paid contractor should be equal to any part of a job, whether writing from scratch, meeting tight deadlines, or dealing with engineers to dig out information. Find out before you sign the contract whether this person can, and agrees to, do what you expect. If he or she can't, or won't, find someone else.

➤ Make your expectations clear and put them in writing. In a hectic situation, it is easy to forget—on both sides—what needs to be done.

➤ Keep your expectations realistic. The contractor is not you, does not know the product as you do, and might not work at your speed. He or she also has less at stake—for the contractor this is just one more project. For you, your job performance is on the line, which affects your raises, bonuses, and promotions down the line.

If you find someone who learns the product quickly and does the job as you want it done, do what you can to keep that person around! If you build a good relationship with a contractor, he or she often will give you "first pick" for his or her available time, which increases the chance you'll get him or her when the need is great. But be aware that the demand for good contractors is high. Providing steady work is a sure way to keep a contractor's attention.

Planning an Entire Project

This, perhaps, is the most daunting task you'll face. It might feel as if you have to be part psychic and part calculator. In reality, project planning has only two dimensions: schedule and scope. Planning a project is a matter of understanding how these dimensions interact.

Considering the Medium

We're not talking about kindness to fortune-tellers, although we believe in that, too. We're talking about the medium that will convey your document content to the reader.

When you plan your document and time, make sure you allow for production time for whatever the medium (or media) will be. It used to go without saying that production time meant printing, binding, and delivering physical books. But now, production applies to anything from those printed books to CDs, HTML pages, and downloadable PDFs. Your planning has to bring everything together at the right time.

For example, say you need printed manuals, the same documentation on CDs, and downloadable PDFs of everything. How long does it take to print, bind, and deliver the number of documents you need? What is production time for the same number of CDs? What's the process for creating PDFs of an entire manual, and how long does it take? These are the processes you have to know about and the time requirements you have to plan for.

Pros Know

"You can never stress too much the need for planning in technical writing. Most writers are anxious to start writing. They feel planning (beyond the bare minimum) is just a waste of time. After all, how can you justify spending three weeks on a project, without having written a single paragraph? On the contrary, the more planning you do, the easier it is to write the document."

—Fabien Vais, Documentation Consultant (Consultant en documentation), Montreal, Quebec

A Bit About Scheduling

Whether you're planning an entire document set and its content, a single document, or content for a corporate Web site or the entire company intranet, you need to understand scheduling. If this makes your head ache, you're not alone. Scheduling is really hard for a lot of people!

Unfortunately, the scheduling "method" most people use seems to be to ask when the project is due and then just cross their fingers and hope it gets done! If your manager asks you when you'll be finished with something, can you give an accurate estimate?

Take heart—there's a better way. The section "Calculating Time" later in this chapter describes a method to replace this madness.

Project Scope: How Wide Is Big? How High Is Up?

The second dimension of project planning is defining what documents are included in the project. Be sure to see Chapter 16, "The Deliverables," for descriptions of documents that might make up your project.

For a Web site, think about what people can do at the site and then define logical divisions for the pages. Will there be a search engine? Archives? Product pages? Troubleshooting help? Are there subdivisions of these? How many total pages might there be? A rough guess in round tens (or hundreds) is good enough.

If you're responsible for planning an entire original document set for a software product but aren't sure what that set should include, think about what the user needs to get started. An installation manual? A tutorial? A quick-start card?

Next, consider what other foundation documents might be needed. What will explain the ins and outs of day-to-day use of the product? A user guide is the thing. How will system administrators know how to maintain and troubleshoot the product? That's what system administrator handbooks and operations references are for. What about adapting the product with custom programming? Developers will look for a product API.

Once you have planned a basic set of documents, prioritize them and start more specific planning on at least the first two. They might be similar ones, such as a user guide and an online tutorial, or they might be at both ends of the spectrum, such as a quick-start guide and a programmer's API. Find out where the need is greatest and start with that.

In most instances, you'll wish you could work on them all at once. But technical writing, like life, often is about compromise. Learn to take the long view: You can always add more titles and more material later, for the next version. Your first goal is to get something—anything—out the door and into people's hands, and to do the best job you can in the time you have.

Calculating Time

Which brings us to that old tyrant, Father Time. When asked to estimate what you can do in the time you have, or how much time you'll need to do a small, medium, large, or extra-large document, you might go completely blank. It's happened to us all, but that doesn't make it any more pleasant when it happens to you.

So blink your eyes, put aside your crystal ball, and try the simple formula shown in this section. It's a tool for getting a rough estimate of how much time you'll need to write a document from scratch. It still requires your judgment, but within an established framework.

Size, Scope, and Quality

There are three factors that affect how long any project will take to be done: the size of a document, the scope of effort required to write it, and the level of quality required. This formula assigns numeric values to different degrees of scope and quality, and combines these with the estimated number of pages to give you a time estimate in work hours.

Size × Scope × Quality = Hours to Complete

Let's take a look at each of these factors.

➤ **Size.** For this factor, plug in the number of pages you need to write for the document. How do you know this number before you've written it? Use similar documents in your company or other companies to give yourself an idea. Is the software more complicated than in the comparison documents, either because of more features or because it's more difficult to use? Increase your page count estimate accordingly. Does your company's product seem 20 percent harder? Increase the page count by that percentage.

➤ **Scope.** Writing a document from scratch takes a certain amount of time to actually write the content, plus many hours spent away from the keyboard and doing tasks other than composing text. You need to hold reviews, type in reviewer feedback, prepare tables of contents, and a host of other chores. In the formula, use whichever of these numbers that corresponds most closely with your situation:

> 1 = Fairly straightforward product that you are familiar with; specifications exist and the developers are available.

> 1.5 = More complicated product that requires learning time, possibly one draft-and-review cycle.

> 2 = Complicated product with no specifications; requires interviews and more than one draft-and-review cycle.

> 2.5 = New product with changes occurring on-the-fly; requires an unknown, but high, number of draft-and-review cycles with many changes each time.

➤ **Quality.** The number you assign for the quality factor is a way to quantify some of the quality measurements we discussed earlier. Expect to work most often with 2.5 to 3.5. (The nature of technical documents makes accurate content a given and non-negotiable element, so we don't include that in this list.)

> 2 = Information must be complete.

> 2.5 = The preceding, plus the text has been spell-checked and proofread, and conforms to the appropriate style guide. More than the barest information has been added.

> 3 = The preceding, plus the document is indexed and edited. Additional information has been incorporated.

> 3.5 = The preceding, plus the document is illustrated with drawings and graphic images. Some effort has gone into producing task-based procedures.

4 = The preceding, plus the document's layout has been designed for maximum usability (can include flow charts, full attention paid to user tasks, special printing methods to add to quality), and the document has been usability tested.

➤ **Hours to complete.** Ah, the magic number! Let's assume your document is 100 pages, the scope of effort is 1 (least complicated), and the quality level is 4 (highest). Using the formula, $100 \times 1 \times 4 = 400$ hours. If the scope of effort increases to 2.5 (most complicated) and the quality level stays the same at 4, the document can take up to 10 hours per page, or 1,000 hours. Quite a difference!

The only hitch in this formula is deciding which numbers to assign for scope and quality and estimating a probable page count. But those will come with experience, and you can err on the high side until then. Your mileage will vary, of course, and you might take much more or less time to produce the job. Or you might find yourself expecting to do a level-one job and as it progresses, find yourself doing a level-four job. We admit this is a rough tool, but it shows how these three factors interact and gives you a place to start.

Pros Know

Industry standards give the following estimates for writing new documents, first draft through final version:

    10 hrs/page for the highest quality

    7 hrs/page for documents people "need and use"

    3 to 6 hrs/page for minimally usable documentation

If you work in an Internet-driven company, however, you might find yourself writing at the rate of two hours per page! So let yourself off the hook a bit if you don't reach the "highest quality" mark in that amount of writing time.

Working Backward: Planning the Schedule

When the future feels like it's rushing at you at breakneck speed already, it might seem reckless to leap out even farther. But that's exactly what you'll need to do to plan your schedule now that you have an idea how long things will take.

To plan a schedule, simply work backward from your target deadline. On a calendar, find your deadline, then calculate the amount of time it will take to do everything that needs to be done between your delivery date and today. To begin, revisit the chronological milestones required for your document. For example, your document milestones, in normal order, are probably as follows:

1. First draft
2. Review
3. Second draft
4. Review
5. Final version distributed to reviewers
6. Camera-ready version completed and handed off to printer
7. Electronic documents posted to the Internet
8. Printed documents in hands of customers

Now, instead of considering them in their chronological order, look at them backward:

8. Printed documents in hands of customers
7. Electronic documents posted to the Internet
6. Camera-ready version completed and handed off to printer
5. Final version distributed to reviewers
4. Review
3. Second draft
2. Review
1. First draft

Begin at the top of the list, and work your way down (back) to the first milestone. Remember, you're working backward from the future (when the document is finished) toward the present (when the document still must be written).

Suppose your deadline is 12 weeks from today. Your thought process will go something like this as you start at the top of the list and work down:

8. Milestone 8, printed documents, must happen in 12 weeks. That date is fixed.
7. Milestone 7 is electronic documents. That takes me less than a week to prepare. I can work on them after I send files to the printer. Milestone 7 will be completed in 11 weeks.
6. It takes 10 workdays (or two weeks) for the printer to finish printing and binding my manuals. Therefore, Milestone 6 must happen 10 weeks from today (one week before Milestone 7).
5. Sometimes the final version goes through changes at the last minute. I have to allow five workdays (one week) for this and to prepare the camera-ready version

of the document. Therefore, Milestone 5 must be ready in nine weeks (two weeks before Milestone 7).

4. After a review, it takes me 10 workdays (two weeks) or more to put in all the changes and send out a new version. But this will be the second review, so changes should be less complicated than the first time around. I'll calculate five days (one week) and hope it is enough. Therefore, Milestone 4 must occur in eight weeks (three weeks before Milestone 7).

3. When I send out the second draft, the reviewers need time to read it and give feedback. Because this is the second draft and I'll mark changes in the files with revision bars, it won't take as long as the first one did. I'll give them three days because they know I'm in a hurry. Milestone 3 must happen in seven and a half weeks (three and a half weeks before Milestone 7).

2. After the first review of a new document, it takes me at least 10 days (two weeks) to put in changes, rewrite according to the feedback I've received, and add all the new material I didn't put in earlier. I'll give myself 12 days (approximately two and a half weeks) because I know I'll need it all. Milestone 2 must happen five weeks from now (six weeks before Milestone 7).

1. The reviewers need a week to read the first draft. I can spend that week indexing, editing, adding graphics, and checking on some of the material I was missing before. Therefore, Milestone 1 must occur four weeks from now. That gives me four weeks, or approximately 120 hours, to write the first draft of this document.

The following figure shows the process as a flowchart.

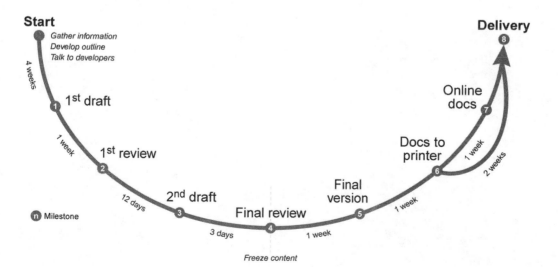

Planning your schedule from start to finish.

(Drawing by Aaron Lyon)

And that's how you plan a schedule. You see now why those milestones we talked about at the beginning of the chapter were so important, and why it can be worth the time it takes to write up a thorough document plan. It's impossible to juggle your time and effort in manageable chunks if you don't know what your milestones are or how long you expect it to take to reach each one.

We flew through the subject of project planning in record speed, and you would do well to follow up by reading at least one good book about the topic (check Appendix B, "For Your Bookshelf"). But you made it! The planning is completed, and now it's time to start writing. From this point forward, your focus is on creating the documents themselves.

The Least You Need to Know

➤ Figure out which is the least important—quality, speed, or price—then focus on the other two.

➤ Turn to another task, rather than an unproductive activity, when you need a break from a main task.

➤ Create a document plan and use it to work backward from your deadline to figure out your work schedule.

➤ Be proud when you learn how to make and meet a schedule—you'll impress yourself as well as others!

Chapter 18

Style Guides: Not Just a Fashion Statement

> **In This Chapter**
>
> ➤ Reasons to use a style guide
>
> ➤ Some examples of real (and avoidable) confusion in writing
>
> ➤ A list of style guides you should own

We've already talked about the software tools you need to know to equip yourself as a tech writer. Those tools are important for forging your documentation. But crafting your documentation takes something more—it takes style.

We aren't talking about how you wear your hat or whether there's a designer logo on your briefcase. Style in writing means doing things in a particular, deliberate way according to set guidelines that govern everything from abbreviations to the correct use of a zero. The collections of these rules of style are called, not surprisingly, style guides.

Is there more than one right way to write? Of course. In this chapter we'll polish up your knowledge of style guides and even tell you how to write your own.

What Is a Style Guide and What Does It Do?

When a tech writer begins to work for a new employer or in a new department, it's not unusual for him or her to ask for the company or department style guide before asking where to get legal pads and sticky notes. Even writers only temporarily assigned to a project usually ask for a style guide while sipping their first cup of coffee. What makes a style guide such a hot property?

In short, it's because the style guide, if it's done well, has the answers to nearly all the writer's questions about how writing needs to be done in that department or that company. Once a writer has reviewed and more or less mastered the style guide, he or she can concentrate on content.

The term "style" refers to two different aspects of writing: mechanics such as word choice, phrase use, punctuation, spelling, and other conventions (often lumped together under the heading "usage"); and more abstract issues such as tone and structure. A style guide addresses both aspects, along with others that are specific to a type of writing.

By setting clear standards, a style guide allows writers to maintain consistency within individual documents and across groups of documents. Consistency, as you'll see later, is essential to effective technical communication.

A style guide presents guidelines for many topics, usually including ...

➤ Abbreviations and acronyms.

➤ Capitalization.

➤ Layout and production.

➤ Product names and corporate terms (for internal guides).

➤ Punctuation.

➤ Spelling.

➤ Tone.

➤ Typographical conventions.

➤ Word choice and usages.

We'll go into all of these areas at the end of this chapter, when we discuss how to create a style guide.

Pros Know

As you work with a particular style guide, you'll find its guidelines becoming second nature. But don't become too set in your ways. What's gospel in one style guide might be heresy in another. There's no single correct style guide—only the one your company or department has chosen to follow.

Style and Speed

At first it might seem as if using a style guide slows you down ("Oh no, I have to look something up *again!*") but the truth is that once you learn the aspects you use most often, you actually can work faster because of it.

One of the interesting—and frustrating—features of English is that it offers many ways to express essentially the same idea, and even many words that mean almost the same thing despite their shadings and nuances. Although this gives English great richness and texture, as a technical writer you'll find that this wealth of options slows you down.

A novelist wrestling the muse has all day to contemplate the virtues of one word over another, but for a busy technical writer, the clock is ticking. Do you really want to spend even 10 seconds of your time deciding whether to hyphenate "end user" in this usage or that usage, or whether "backup" needs to be one word, two words, or hyphenated? How much more efficient you'd be if you could focus instead on content—your company's product!

Where a style guide helps you pick up speed is in those precious seconds you spend every time you pause to decide which term to use, whether to hyphenate, and whether you should say "Click X to do this" or "Click on the X button to do this." This effect is magnified by the guidance you receive on overall issues such as tone and how to address certain topics, if the style guide covers those (and many do).

If you work in a department of several writers or with an editor, this times savings is multiplied yet again. For each minute you save yourself by writing the correct term without thinking, or by structuring procedures correctly the first time, you've also saved the time an editor or colleague (or both) would have had to spend identifying and correcting your inconsistencies.

Guideline or Requirement?

A style guide is meant to be just that—a guide, not a book of hard-and-fast rules. Like life, tech writing is full of the unexpected, and there will be times when you have to go it on your own—against the style guide's rules.

When is it okay to ignore the rules? Just as in life— when you have a darn good reason. Will clarity be lost or common sense violated if you blindly follow the style guide? If so, that's a good reason. Is a particular writing circumstance unique to this section, this document, or this product? That's a good reason, too.

Ultimately it is still up to the writer's best judgment about what to do to communicate clearly and accurately. So in the end, the rule is that "you rule." Your readers depend on your judgment, ability, and plain common sense. Remember, technical writing is about communicating clearly, not blindly following rules.

Dodging Bullets

If it takes you twice as long to figure out how to write something to match a guideline, it's probably an instance in which you can safely ignore it. Make sure the juice is worth the squeeze.

Pros Know

"Technical writing is, among other things, writing that does not draw attention to itself as language, whereas poetry and fiction are always drawing the reader's attention to the language as language. People who consider themselves primarily poets or fiction writers often have a hard time writing prose that conveys information to the reader in writing that is as unobtrusive as possible."

—Michael Leonard, Group Manager, AT&T Labs Technical Information Design and Development

Why Consistency Is Important

Consistent word use is vital in technical writing because it gives the reader a firm handle by which to grasp concepts, processes, and instructions. There's no question of what's what, who is who, or when one action is the same as another.

Not only is it good practice to avoid using different words to mean the same thing, but also avoid using the same word to mean different things. For example, using "select" to mean both "click on" and "choose in your mind" is sure to cause confusion.

Many words common in the computer industry are similarly tricky. Here are three: "activate," "enable," and "specify." These sound deceptively like nice, vigorous, technical verbs—but what exactly do they mean? What does it look like when a user is "activating" a feature or "enabling" an option? When you tell a user to "specify the circuit" does that mean to click an icon representing the circuit, to set an actual switch somewhere, or both—or something else?

Dodging Bullets

If you feel you might be using a term incorrectly, it might mean you need to learn more about your topic—or that colleagues aren't using the term consistently themselves. Consult a technical dictionary for an industry term, or your company's glossary if it's product- or process-related. If you're still not clear, ask two or three of the developers in your company (and let them know if their explanations don't match!).

But it's not impossible to use these words effectively. One way is to define the word at its first occurrence in a chapter or section. Another way is to follow the word immediately with a description of what the action consists of: "Activate the XYZ feature by clicking XYZ in the Options menu."

The language of software and hardware is particularly guilty of "word recycling"—using the same term for different things at different times. For example, Company A develops an E-Widget that backs up your files, but later, Company B's E-Widget erases your hard drive.

Another term entering the "recycling process" appeared in our earlier example: "token." This term, formerly limited to board games and subways, now can be used to mean an element of a programming language, a series of bits that allow a message to be sent over a network, and (in security systems) a device with a secure password and ID that lets a user log in to a network—sometimes all in the same document. And it still applies to board games and subways, too!

Of course, you can't single-handedly prevent the repeated or inaccurate use of industry terms or acronyms at your company. You can, however, do your best to be aware of the problem and make sure to clearly define these potential troublemakers. As quickly as new products and processes take shape at your company, abbreviations and acronyms will pop up to refer to them—not to mention pre-release code names. All too often these are already in use, if not at your own company, certainly at another. If there's even a small chance a reader could be confused, spell the term out (at least once).

There also must be consistency in factors of formatting such as how different type styles are used. For example, perhaps you like to use italics to identify variables in messages or user input. But another writer in your group likes to put variables in brackets. These kinds of differences are bad enough from one document to another, but when they occur within the same document, it's nothing less than disastrous. Readers assume differences have meaning—so make sure they do.

This extends to all other formatting aspects as well, from the way steps in procedures are structured to the elements included in section or chapter headers and footers. Consistency sets up expectation in the reader's mind, then meets that expectation. That's not boring, it's reliable.

Follow the Bouncing Word

In the following sentences, the writer uses three different words to tell the user what to do:

➤ Activate the connection.

➤ Enable the connection.

➤ Turn the connection on.

Do they all mean the same thing? How would you know?

And what about simple terms in instructions such as "*Choose* Close from the File menu," "*Select* Close from the File menu," and "*Click* Close on the File menu." Even with those commands that you think are so familiar to everyone, there's room for doubt. And with good reason. Even though computer familiarity increases every day, it wasn't so long ago that tech writers were explaining what a mouse was and how to use pull-down menus. Those aspects of computer use have become standards, but many other aspects are still up for grabs with each new product.

Coffee Break

Homonyms are words that have the same sound and often the same spelling but different meanings. Homographs are words that have identical spellings but different meanings and often different pronunciations. Heteronyms are words that are spelled identically but have different meanings when pronounced differently. For example when the word "leading" is pronounced *ledding*, it means "in the forefront," but when pronounced *ledding*, it means the distance between lines of type.

Coffee Break

According to Media Metrix, in the year 2000 more than 55 million American households had home computers and there were many times that number in the business world.

Not convinced? In the sentences above, how do you know you'd use a mouse to choose or select "Close" from the File menu? And does "click" mean you *must* use a mouse, or could you do the same thing with a combination of keystrokes? Not quite so obvious after all. A good style guide can help you here, too, because you can expect it to anticipate these kinds of word-usage questions and address them in a clear, definite way.

For Me to Know and You to Find Out

Forcing the reader to play guessing games is never a good idea. There's no need to make the user play a round of "What's My Definition?" as you try out one word and then another to refer to an object or action. This can happen when a tech writer hears something called an internal name or a nickname, or when the tech writer inserts material written by someone else, and the new material refers to an object by a different name. Unsure of what the product name really is or if the new name refers to a new object or the same one, the writer, perhaps with no time to dig for details, tries to cover all possibilities.

Maybe you're writing something about the Chrysalis-ITS Luna CA token, which you know is a type of "Luna token." An engineer in your department gives you material in which he seems to refer to it as a "Luna card." You look in the Chrysalis-ITS documentation but don't see the term "Luna card" anywhere. However, it's so prevalent in your company, you're not sure if it's a real term or not. Then there are the terms "PC card," "smartcard," "PCMCIA card." Are all of these synonyms, or are some of them different products? You think they all mean the same thing, so you make sure to mention all of them, figuring that at least some apply to what you're talking about.

Bad idea! If you don't understand it, it's a guarantee your user won't understand it. And using six different words to mean the same thing can only create confusion. Each time you use a different term, the reader will think (sometimes rightly) that you are introducing a new item.

What if you had to use or install a complex and expensive product all on your own, and the documentation continually used five different names for the same thing? That's exactly the situation countless technicians and administrators find themselves in every day.

Dodging Bullets

Even if a developer hands you written material as "ready to insert," you should still make sure you understand what it says before dropping it into place in your document. We guarantee that the part you don't understand is exactly what someone will later ask you to explain. Remember, if it isn't clear to you, it won't be clear to someone else.

"Elegant Variation" Doesn't Improve Style

The fifteenth time you write the same term for your product, it might be tempting to find another way to refer to it. After all, you're bored with writing the same term over and over—won't your reader be bored with reading it?

In a word, no. Readers who are focused on following important processes or understanding new concepts aren't counting the number of times a given term appears. In fact, switching to a different term just to make things more "interesting" is more likely to trip the reader up.

In other words, don't employ "elegant variation"—using a different term solely for the sake of novelty—in a technical document. When you introduce a new term, or change from one term to another, have a good reason, and make that reason clear to the reader.

Classic Styles

Some style guides have become standards over many years; others are new entries tailored for the computer industry. Here's a list of some style guides widely in use in the United States and Canada. All would be valuable additions to your bookshelf.

Coffee Break

What exactly is "elegant variation"? We like this vintage explanation excerpted from H. W. Fowler's *The King's English, 2nd ed.,* 1908, Chapter 3, "Airs and Graces": "... Many writers of the present day abound in types of variation that are not justified by expediency, and have consequently the air of cheap ornament ... the variation is often worse, because more noticeable, than the monotony it is designed to avoid."

➤ *The Chicago Manual of Style.* Since 1906, this vintage volume has been an all-inclusive guideline for writers and publishers in all fields. It addresses punctuation, documentation, foreign languages, indexes, design, typography, and more. We recommend a hardback edition—it's sure to see long and hard wear.

➤ *Microsoft Manual of Style for Technical Publications.* If you could have only one book as your tech writing style guide, this might be the one. It contains the styles Microsoft follows in its own documents, valuable knowledge in view of Microsoft's domination of the industry. It also tells how to use and spell both general and computer-related terms as well as defining acronyms and abbreviations.

➤ *Wired Style: Principles of English Usage in the Digital Age.* More than a resource for general writing or even general technical writing, this book is aimed toward high-tech communications writers. It contains guidelines for Internet style, a discussion on writing technical material, a dictionary of relevant terms, and a style *FAQ* (a list of *Frequently Asked Questions*).

Pros Know

FAQ is an acronym (pronounced *fak*) or abbreviation (pronounced *eff-ay-kyew*) for **Frequently Asked Questions;** it refers to a list of questions commonly asked by people new to a subject and includes answers to those questions. FAQs originated on Internet newsgroups and now often are part of Internet sites and mailing lists as well.

➤ *Read Me First: A Style Guide for the Computer Industry*. This is Sun Microsystems's style manual, addressing everything from basic punctuation and style pointers to legal guidelines, writing for an international audience, and creating a documentation department.

Of course there are other published style guides tailored for writing in different fields. The *U.S. Government Printing Office Style Manual,* for example, is an invaluable aid if you work for the government. Newspaper reporters probably will be most familiar with the *Associated Press Style Guide,* which focuses specifically on journalism.

If you plan to specialize in a particular sector of the computer industry, you might find special style guides that apply. If a Net search doesn't turn up anything, ask someone who writes in the area you're interested in.

We recommend adopting one of the major guides as your foundation style guide if your department doesn't specify one for you. If your company's product runs on or is associated with one of the Windows operating systems, we recommend you adopt *Microsoft Manual of Style for Technical Publications* because your readers will be familiar with the Microsoft style.

If, on the other hand, your product runs on or is associated with UNIX or Linux, you might do better with Sun's *Read Me First*. Compare the two and choose whichever one seems better for your circumstances.

Creating a Style Guide

If your company hasn't chosen a style guide to follow, think twice before you dive into developing a full-blown style guide yourself. There's no point in reinventing the wheel—the publisher's staff for an existing style guide have already put in months or years on this very task, so what's the point in you starting today from square one?

Technical writers in the computer industry usually adopt one of the big-name style guides as a foundation for basic style decisions. It's then common practice to supplement that tome with a lean, focused style guide to address topics specific to the company, or even to a particular line of products within a company's family of offerings.

The Mini Style Guide

Even if your company doesn't have an official style guide, and there's no mandate to create one, there's no reason not to go ahead and make your own, even if it's just a "bare bones" version. The absence of a style guide doesn't mean you don't need one. On the contrary, it probably means you need one more than ever.

Here's how to begin your style guide: As you start to write or update a document, make notes of how you spell or capitalize a word, how you format elements such as bullet lists, and how you punctuate. Keep track of the words you're unfamiliar with and how they're used (you also can build a company-specific glossary as part of your style guide). If you see a word used inconsistently, see how it's used in other company-published documents, if there are any. If you capitalize a word or phrase once, your notes will help you make sure you capitalize the same term the next time you have to write it. If you're truly breaking new ground and there aren't other documents to compare, just decide on the usage that makes most sense to you and make a note of it so you will stick with it.

At first, your modest guide will barely fill a single sheet of paper and it will be useful only for the document you're working on. But make it readily available to others and before long, you'll probably see your guidelines becoming accepted company practice. It's time to publish it officially then, because you can be sure other people in the company will want something to replace the "bootlegged" copy they've been using!

Just Another Technical Document

Approach writing a style guide just as you would writing any technical document. This means being thorough, clear, accurate, and organized. Gather all information that isn't covered in your basic style guide (we'll use *Microsoft Manual of Style for Technical Publications* as the example here). Easy to say, harder to do. Start by jotting down categories specific to your company, such as your company's product names. It's important to make sure that you, and any new writers coming to the department, spell all product names the same way—and the right way!

Keep a notebook or computer file devoted to this subject. Pay attention to the times when you pause to decide style issues such as how to spell a word, or stop to try to

Pros Know

See that your style guide devotes a whole section to correct spelling and usage of company-specific words and terms. You'll be amazed (and not in a good way) at how often people misspell product names. Be sure to keep the information up to date.

Pros Know

When creating a style guide, group your topics into no more than two or three main sections, such as "Word Usage and Grammar," "Formatting," and "Production Issues." Then, alphabetize the subsections. Organizing by the alphabet provides a fixed, nonarbitrary order that everyone knows how to use.

remember what typographical format or term is used, and make a note of the word or term. Also note whether you found an answer in your foundation style guide or had to decide on the answer for yourself. (We're still not reinventing that wheel.)

When you feel you have reached "critical mass," create a draft of the guide. Be sure to share it with any other writers you work with—they're sure to have their own valuable ideas to contribute. Like any other technical document, developing your style guide calls for getting knowledgeable feedback and building stakeholders.

How Do You Know Which Style Is "Right"?

Unfortunately, there is no style guide that's always "right." All style guides are, in the end, only expressions of opinion. Rather than trying to figure out which one is "right," we suggest focusing instead on which one gives you the best results—documents that are clear and readable, and that follow conventions your readers are familiar with.

We'll never have a divine decree to settle whether "e-mail" should or shouldn't have a hyphen (although some writers write as if they believe they *have* received a divine mandate). All we can do is emphasize that having a consistent style is more important than which style guide you follow.

When all's said and done, no matter how many style guides you've read or guidelines you've memorized, you still have to fall back on your own good judgment. But style guides help you feel confident that your judgement is informed and sound, and that you can explain it to someone else if you need to.

The Least You Need to Know

➤ Familiarize yourself with the standard style guides, and choose two or even three that complement each other.

➤ Save writing time and produce better documents by addressing company-specific style issues in a supplementary custom style guide (write it yourself if you have to).

➤ Be consistent in how you address style issues, regardless of which style guide you choose to follow.

➤ Remember the ultimate value of your own good judgment about what is clear and readable—style guides provide only guidelines, not unbreakable rules.

Chapter 19

Writing Clearly

In This Chapter

➤ Practices to make your writing shine

➤ The importance of simplicity

➤ Procedures are as simple as 1, 2, 3

➤ Considerations for an international audience

Clear writing—writing that doesn't make the reader guess what a sentence or term means, or how ideas relate—is where the rubber meets the road.

But being able to write clearly doesn't take long years of monkish discipline or top-secret neuroprogramming. It is a learned skill that's based on some simple (and clear!) guidelines, as well as down-to-earth common sense.

If writing isn't your strongest point, this chapter will help you breathe more easily. If you already feel confident as a writer, this chapter will show you how to tune your writing for technical documents.

But even these rules, helpful as they are, shouldn't be followed blindly. Although it might be easiest to learn and follow a set of rules, this isn't always possible in tech writing. What we describe in this book as being the "right way" or even the "recommended way" is based on our many years of experience in the field and our knowledge of how tech writing is done around the world. But, as you know, rules are made to be broken. The bottom line is not how well you can follow the rules but how well the user understands what you write. If following a rule (including any of those you find in this book) means you have to write an awkward or difficult-to-understand sentence, don't do it. If following a rule means you have to spend a huge amount of time figuring out a way to follow that rule, don't do it. The return on the investment just isn't worth it.

Best Writing Practices

No, we aren't talking about writing "I will be clear" a hundred times. By "writing practices" we mean the guiding principles—or practices—that govern good technical writing. These can become as automatic for you as breathing and walking and eating chocolate—well, breathing and walking anyway.

Pros Know

"Grammar isn't a set of rules. It's a snapshot of how a language is used at a particular place and time. Proper grammar is simply a snapshot that's twenty or thirty years old. Use grammar as a guideline. If you can write more clearly for ignoring it, then ignore it."

—Bruce Byfield, Outlaw Communications Contributing Editor, *Maximum Linux*

Respect the Reader

At the core of most of these practices is a single, essential kernel: respect for the reader. If you respect the reader, you don't feel the need to tap-dance around facts or fancy up phrases with today's entry at www. bigwords.com. Speak to the reader as if he or she were sitting next to you, asking for help. And that's exactly the situation when the user turns not to you but to the document you've created.

Here are the best of the best—the practices you should eat, drink, and sleep with under your pillow. Like in a good style guide, we've put the topics in alphabetical order.

Active Voice: Don't Worry, Be Active

Active voice (not the same as imperative voice—see the "Imperatives: That's an Order" section later in this chapter) is one of the cornerstones of clear writing. Voice is about the relationship between a sentence's subject and its verb. (Recall from high school English that the subject is what the sentence is about, usually a thing; the verb is the action word *about* the subject.)

Not surprisingly, the opposite of active voice is passive voice. Trust us, you'll find yourself in discussions about both. So you need to see why active is nearly always better, but also recognize when passive voice is called for.

In active voice, the subject *does* the action of the verb. For example, Jim installs software. The sentence is about Jim, so he's the subject. What's he doing? Installing software.

In passive voice, the subject *receives* the action of the verb, and it might not be clear who or what actually does the acting. The subject just lays there passively and is acted upon. For example, "The network card was installed." The sentence is about the network card, so it's the subject. "Installed" is still the verb. But who did the installing? There's no way to know.

That's the greatest weakness of passive voice—what we like to call "the phantom effect." Things seem to happen as if by magic. Software is installed. Options are chosen. Networks are configured. Files are copied. Aside from being unclear, it gets downright spooky.

Still not sure you can spot passive voice? Look for some form of the verb "to be" (am, are, is, was, were, be, been, being) coupled with the past tense of another verb (usually ending in -ed). Take another look at the previous paragraph to see some examples. This construction is the red flag for passive voice.

Active voice, on the other hand, puts everything right out on the table in daylight. Relationships are clear. The customer installed the software. System engineers configured the networks. A programmer copied the files. There's no "man behind the curtain" pulling levers and pushing buttons that we can't see.

Another strength of writing in active voice is that it forces you to understand how things get done. Before you can explain it to someone else (in active voice, of course), you must understand it yourself. If you're tempted to write a sentence using passive voice because you really don't know where an action is coming from, that's a cue you need to learn more about your topic.

But passive voice is not universally evil. It can be useful and appropriate in the right situations. Because the subject receives the action of the verb, you can use passive voice effectively when you need to emphasize the subject and the result of the action, rather than the actor. For example, maybe it doesn't matter what part of a system updates a set of database tables, only that it does so on a specific schedule. In that case, writing "The database tables are updated every 15 minutes" is perfectly fine.

Pros Know

Active voice is powerful in types of writing other than technical documents. Using active voice makes e-mail and memos more direct, and increases credibility. In speech, an intention to use active voice helps you think more clearly and speak with greater confidence.

Dodging Bullets

There's little room for "always" and "never" in technical writing, and the issue of active voice is no exception. Beginning tech writers sometimes work to make every sentence active voice—and end up with paragraphs that only sound forced. When you can use active voice in a natural way, do it—but don't struggle for it. The result is seldom worth it.

Just don't use passive voice to cover up a lack of knowledge or to misdirect the reader's attention. You're sure to encounter passive voice used for those very things in all sorts of documents—don't make that mistake yourself.

Commas: Dangerous Punctuation—The Serial Comma

What is a serial comma? It's a comma used to set off the individual elements in a series of elements in a sentence. For example, in the sentence "After winning the lottery, Jackie vacationed in Tahiti, Bali, Fiji, and Tasmania" the three location names that appear in a row (Tahiti, Bali, and Fiji) are the series elements. For clarity, always include a comma after the last item in the series. In the example, that's the comma following Fiji.

Dodging Bullets

Punctuation is powerful—make sure you use it correctly. Too many tech writers—along with developers, designers, programmers, and engineers—turn up their noses when it comes to considering punctuation. Yet those little dots and marks are essential to any sentence, and misplacing one or more can radically skew a sentence's meaning. You aren't being prissy when you worry about punctuation—you're being precise. We don't mean you should obsess about it, but you should be able to explain why you used every punctuation mark in your document. (Someone actually might ask you to!)

Leaving out the last comma in a series is dangerous indeed because it can lead to serious ambiguity for the reader. Consider this sentence: "Behind these doors are money, a lady and a tiger." Does this mean the lady and the tiger are together behind one door? (We hope not.) And how many doors are there—two or three? Don't leave your reader vulnerable to doubt. Always keep that last serial comma where you—and the reader—can see it.

Emphasis: We Really Mean It

There will be times when a word or phrase deserves special emphasis to attract the reader's attention. Perhaps it's a note about sequence, or the consequences of taking an action. Whatever the cause, it's vital that the reader knows to pay special attention to those words.

With so many options available in DTPs for adding emphasis—bold, italics, bold italics, color, underlining, or even all these combined—it's easy to use it as an excuse to express your creativity. Or perhaps it's tempting to use different kinds of emphasis to indicate different kinds of importance—bold for warnings, italics for important notes, underlining for critical steps in a process. Don't do it. It's much better to choose a single method for adding emphasis and then to use it sparingly. Why?

First, emphasis adds a visual element that the reader must interpret. It might be clear that the emphasized text is different from the words that surround it, but what does that difference mean? The reader must figure out what meaning to assign to each form of difference. (Even if the document includes a list of conventions, the reader must still remember them.) This all adds effort.

Second, with a document such as an administrator's handbook that discusses scripts, files, directory structures, and commands (sometimes with placeholders), you're likely to need many of those methods of adding emphasis to differentiate these for the reader. Perhaps file names and directory structures always use italics, or commands appear in a particular font. With this much visual complexity, it's not enough for a word simply to be different to show emphasis—it must be different in a consistent and particular way.

Last, emphasis is like speaking emphatically—who likes to listen to somebody who always talks too loudly? When emphasis is overdone, readers begin to ignore it. And that's hardly what emphasis is all about.

Gender-Neutral Language: He Said, She Said

English has the peculiarity of having a gender-neutral pronoun for groups of people—"them" or "they"—but not for individuals. For that we're stuck with "he" or "she." (Nobody so far has voted for being called "it.")

No doubt you'll hear the argument that using "he" to mean a reader of either gender is acceptable because of historic convention. Another common tech writer debate. Our reply is that times change and so does acceptable usage, especially when clarity and plain accuracy are at stake. Always using "he" is no more clear or accurate than always using "she." So unless you are certain that one or the other is accurate, avoid using a one-gendered pronoun.

But what to do instead? Writing "he or she" again and again can be tiresome for the writer and distracting for the reader. The construction "s/he" has gained acceptance in some areas, and it does have some points in its favor, but many people still find it unsatisfactory.

We have also begun seeing "they" and "them" used in constructions such as "When the user enrolls for a certificate, they have to provide their password," but the obvious number disagreement here ("them" is plural, but there's only one system administrator) can contribute to misunderstanding. (However, this style is becoming more common and might be done in your department.)

Don't despair—there are several tactics that work well. One of the most effective tactics is simply to "write around" the issue. This means constructing your sentences in ways that don't require you to say "him or her." It isn't as hard as you might think.

215

Dodging Bullets

Three words to watch out for in technical writing are *let, enable,* and *allow.* Although *let* seems to have the advantage of being short and direct (usually a good thing in tech writing), many people read it as having to do with giving permission—seldom the case in a technical document. *Allow* can have a similar problem, depending on the sentence it's used in. *Enable* is a bit trickier—frequently it means to "confer ability" but sometimes it is program-specific, usually about activating a software feature or hardware capability. Knowing your product should help you avoid confusion; if your audience and your product use "enable" in this technical sense, then you should do the same, and use *allow* when "conferring ability."

For example, instead of "Ask a sales rep if he can give you the latest marketing materials," write, "Ask a sales rep to give you the latest marketing materials." Instead of, "The user must provide their password," write, "The user must provide the password." Or take advantage of the second person and avoid all other pronouns by writing, "You must provide your password."

Make the subject plural so you can use "they" or "them" legitimately. Instead of, "If a user is confused, they should read about error messages," write, "If users are confused, they should read about error messages."

Save "he or she" for times when there really is no other good alternative—then use it without guilt. Occasional use won't trip up the reader. The most challenging part of nonsexist writing is simply raising your own awareness about it. Once that's done, incorporating it becomes easier and more natural.

Humor: Funny Is as Funny Does

Humor in tech writing is a funny thing. No wait, that's not what we meant. And there's exactly the problem with trying to use humor—it seldom comes across the way you mean it.

A sense of what's funny is highly subjective, and varies widely from person to person and from culture to culture. Think of your favorite jokes—the humor in them seems obvious and irresistible (to you). Then think about jokes you don't get the point of or are even offended by. Yet somebody else thinks they *are* funny. There's just no accounting for taste.

Another problem is that humor in writing is different from humor shared in person. Tones of voice, facial expressions, even gestures and postures make up a lot of a joke's overall impact, and those are all missing from the written word. But they're essential to clearly communicating mood and intent—that's why *emoticons* were invented for e-mail, newsgroup postings, and chat rooms where the written word stands alone.

Humor is rather boring, too, the tenth or twelfth time you have to read it or listen to it. Technical documentation is rarely meant to be used only once, so don't do anything that might make your user grow tired of or annoyed with the document.

The bottom line is that humor doesn't do a thing to improve clear, solid writing, which after all is the primary objective. So although we applaud your desire to entertain your readers, we have to say it's generally a mistake. Take our advice—please.

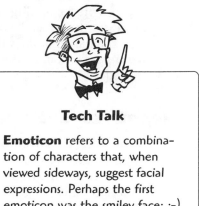

Tech Talk

Emoticon refers to a combination of characters that, when viewed sideways, suggest facial expressions. Perhaps the first emoticon was the smiley face: :-) Tilt your left ear toward your left shoulder and you'll see the little face. Emoticons give writers a way to include missing elements such as tone of voice or a playful mood in what they write.

Imperatives: That's an Order

When you're writing instructions, use the imperative voice (not the same as active voice—see the "Active Voice: Don't Worry, Be Active" section earlier in this chapter). Imperative voice means using sentence structures and verb forms that tell someone to do something in no uncertain terms.

Imperative voice is different from active voice because imperative voice always tells someone to do something—press, install, read, type, launch, open, close. Active voice simply makes a statement and gives no directive to the reader: "The CD-ROM provides documentation and code samples."

In imperative voice, it is always understood that the writer is addressing the reader in the second person (you). The imperative sentence structure looks like this: "(You) imperative verb." For example, "(You) stand up." It also might look like this: "(You) imperative verb object." For example, "(You) press the Shift key." It might help to think of it as "Hey you" instead of just "you."

Inexperienced writers shy away from using the imperative voice because they are afraid of sounding rude or bossy. Far from it! Using the imperative voice shows your confidence and helps the reader trust you. A sentence such as "Type your name in the Name field and press Enter" tells the reader exactly what to do and leaves no room for misunderstanding. Your reader will sigh with relief.

If you write something like "Your name goes in the Name field and pressing Enter processes the screen," you've left the reader with questions. Does he or she need to type the name, or does the system fill it in? When's the right time to press Enter? Feeling such doubts, the reader starts to wonder if the instructions can really be trusted.

Negatives: Let's Be Positive, People

Avoid using negative words in tech writing. (Just like that: Notice we didn't say, "Do not use negative words.") There are two reasons for this guideline. First, it makes writing less clear. One negative often leads to a second, and it can make it very difficult for a non-native English speaker to ferret out the real meaning. Try to understand this sentence:

"The Close button is not enabled if you have not selected the correct item."

Two negatives in one sentence, and we still are not sure what the writer meant to say.

Using negative contractions is doubly difficult for translation purposes or for English-as-a-second-language readers, especially if the contraction also spells a word: *Can't* might be misread as *cant*; *won't* might be read as *wont*.

Second, negatives suggest that something's wrong, and manufacturers don't like to have their products referred to in any negative terms!

Tech Talk

Bridging word or phrase (also called "transitional" or "linking") is a word or phrase that indicates a connection or other relationship between two or more paragraphs. Examples of bridging words and phrases are "additionally," "on one hand/on the other hand," "consequently," and "meanwhile."

Paragraph Length: Six Is Enough

An important part of writing clearly is not overloading the reader in any one paragraph. In technical writing, small paragraphs—even single-sentence paragraphs now and then—make it easier for the reader to take in the core idea in each one. This is especially important when addressing highly technical topics or explaining complicated processes or procedures.

A good rule of thumb is to keep paragraphs to no more than six lines if at all possible. Not six sentences—six *lines*. If you can't get the paragraph's single clear idea across in six or fewer lines, that's a strong clue you need to look more closely at what you're trying to say. You probably have more than one idea.

If so, split up the paragraph into two smaller paragraphs and use a *bridging word or phrase*, such as "additionally," to start the second paragraph and tie the two together.

Of course, there will be times when six won't be enough, and that's okay—just treat it as a guideline or benchmark, or a tool to keep the writing focused, not a hard-and-fast rule. (If you're paying attention, you'll notice we exceed the six lines guideline in this book. That's because this isn't a technical document—it's a mass-market book written according to different guidelines. We're paying attention, too!)

Parallel Structure: No Surprises

Along with all the other things, keeping writing clear means telling the reader what to expect and then delivering on that expectation. Do this by keeping structures parallel at all levels within a document.

What are structures, and what's it mean to keep them parallel? Structures support the actual text—such as how chapters are organized internally, or the order in which elements in a section or even a paragraph or sentence are presented.

For example, a chapter structure might be to introduce a topic, write step-by-step instructions about it, then close with a list of limitations or parameter values. The reader begins to expect to see that in each chapter about similar topics. When such structures are repeated from one chapter to another—that is, the chapter structures are parallel—the reader can focus on the content instead of spending time discovering how the information is organized. You can see parallel structure at work in the book you're reading.

Parallel structure works at all levels of a document, not just on the large scale of chapters or sections. For example, the bulleted list provides an opportunity for parallel structure in the way each bulleted item is expressed. Use the same structure for each item—you might start each with a gerund (-ing), make each a complete sentence, or make each a completion of the sentence that led into the list. Just do the same thing with each item.

Present Tense: Be Here Now

By that we mean, "Write in the present tense." Because the reader executes actions at a point in time later than the actual time the writer writes them, beginning tech writers fall into the trap of writing in the future tense: "When you press the Enter key, the system will display the next screen."

What's wrong with this? The problem is that "will" raises that old specter, ambiguity. For example, if you write, "When you complete this process, the system will update the database," do you mean updating happens immediately, or do you mean that the database is now in a state so that the system can update it at some unknown, later point in time? In that earlier sentence, does the system actually display the next screen immediately, or has it simply been rendered *capable* of displaying the next screen? Ambiguity, grinning its terrible grin (or is it a snarl? we can't tell) ....

If, on the other hand, you write, "When you complete this process, the system updates the database," there's no question what's going on and when it's happening. Likewise with "When you press the Enter key, the system displays the next screen." The reader isn't left wondering when an event "will" happen.

Of course, sometimes you need to indicate that something happens in the future, and in that case by all means use "will." But save it for those times, and indicate an event time frame whenever you can. Careful use of tenses—past, present, and future—keeps readers firmly anchored in the here and now.

Second Person: Yes, You ...

Address the reader directly by using "you." This also is called addressing the reader in the second *person*. The book you're reading now addresses the reader (you!) in the second person. Speaking directly to the reader helps clarity in several ways:

➤ The reader knows the writer is talking to him or her, not making undirected statements or observations.

➤ It makes using the imperative voice (see "Imperatives: That's an Order" the earlier in this chapter) more natural, because when you tell someone to do something, the "you" is understood.

➤ The document takes on a more casual, friendly tone, as if there's just one person speaking or writing to another. This helps readers relax and focus on the content.

Is this informal tone really acceptable for technical writing? You bet. The detached, formal style used for things such as reporting scientific research findings has no place in the computer industry. For one thing, it takes too long to read and understand, and who has the time? For another, the tone of this industry as a whole is more "business casual" than "suit and tie." So relax without guilt.

Tech Talk

Person, *in writing, is a way to indicate how near the reader (or writer) is to what is being said. The first person is "I," "me," or "we;" the second person is "you;" the third person is "he," "she," "they," or "them."*

Don't step in the you-do. This sounds unpleasant—and it certainly is for the reader. A common mistake beginning tech writers make when writing instructions in the second person is to use "you" in a declarative way when what is needed is a directive. This results in the dreaded "you-do"—"First you do this, then you do that," and so forth.

But you-do is easy to clean up—just take out the "you." A sentence such as "To begin installation, you put the CD-ROM in the drive," becomes a clear directive: "To begin installation, put the CD-ROM in the drive." And you don't need to change your shoes.

Should: Should You or Shouldn't You?

You shouldn't. Or, to be clearer, avoid using the word "should." It gives the sense of the writer shaking an admonitory finger at the reader, with not enough clear information for the reader to make an informed decision. Consider this sentence: "You should close all other programs before installing a new one." Well, what happens if we don't? Are there dire consequences awaiting us, or does the writer just think it's a good idea?

When you find yourself wanting to write "should," stop and think about it. Try substituting "must." Is the sentence still true? If it is, write a clear directive: "Close all other programs before installing a new one." Or even "You must close all other programs before installing a new one" if the consequences of leaving other programs open are serious enough.

On the other hand, if substituting "must" makes the sentence untrue, clearly what the reader needs is a recommendation rather than an instruction. Rewrite it as such: "We recommend you close all other programs before installing a new one."

If you're not sure whether a "should" is a "must" or only a recommendation, stop and find out before you write one or the other. It could make the difference between a happy user and one whose whole system just crashed.

KISS—Keep It Short and Simple

With technical subjects that often are complex and need long documents to explain them, it might sound contradictory to say "Keep it short and simple." But it is exactly this fact about topics in the computer industry that make "short and simple" so important in technical writing. Readers have enough complexity to grapple with in the subject itself—they don't need more in the documentation.

Call a Spade a Spade (Not a "Manually Operated Multipurpose Soil Manipulation Instrument")

When noted American author Mark Twain was writing for newspapers and being paid seven cents a word, he made this comment about using short, simple words: "I never write *metropolis* for seven cents because I can get the same price for *city*." Technical writers aren't paid by the word, but the same principle applies—when a short, simple word will get the idea across, don't use a longer one.

Writers who lack confidence usually are the ones tempted to substitute "obtain" for "get" or "utilize" for "use." But technical writing isn't about showing off your education or your vocabulary—it's about communicating a technical subject to a reader who is in a hurry. Shorter, simpler words are, by their nature, less ambiguous—and that's all to the good when clear writing is the goal.

Take this example from a real interoffice memo: "Should personnel in your department require additional assistance, please register a request with a Technical Support representative." What they meant, of course, was simply this: "If your people need more help, contact Technical Support."

If you still aren't convinced that "use" is more readable than "utilize," perhaps the Fogg Index will change your mind. It's a formula for determining readability, expressed as a grade level, based on a 100-word writing sample. (By grade level we mean grade in an American system where 12th grade is the last year in high school.) College texts are written at an 11th–14th grade level, a good level for a technical document. If you are trying to reach a wide audience, aim for a lower grade level; for a specialized programmer's document, a higher level is fine.

To assess the readability of a 100-word passage of your document:

1. Count the number of words with three or more syllables ("long" words).

2. Count the number of sentences in the passage.

3. Use the following formula to determine grade level:

.4 × (number of long words + (100÷number of sentences))

We can apply the formula to the preceding paragraph, which contains 100 words: .4 × (13 + (100÷5)) = 13.2, or 13th grade, a slightly above-average grade level for a college text.

And, with necessarily long terms such as "configuration parameters," there will be plenty of long words in any technical document already. Don't add to them unless there's a good reason to.

Learning from Simplified English

You will think twice when tempted to use a long or obscure word after looking into AECMA Simplified English, a writing standard for aerospace maintenance documentation. Simplified English was developed for non-native English speakers but also has fared well with native speakers. This type of writing standard is known as a controlled language because it restricts grammar, style, and vocabulary to a subset of the English language. Simplified English restricts allowable words to about 950, with each word having only one meaning. Check out www.aecma.org/Publications/SEnglish/senglish.htm for more information.

Dodging Bullets

There's no doubt a good thesaurus can be a great resource. But it also can get you in trouble. Words that appear under a thesaurus entry don't all mean the same thing. Like spices in cooking, they all have their own nuances of flavor and complexity. Substituting "expunge" for "delete" might make your writing seem spicier, but it doesn't make it clearer. Good writers, like good chefs, know that more isn't always better. Use your thesaurus sparingly.

Simple as 1, 2, 3: Writing Procedures

It's as certain as sunrise that at some point you'll be asked to write procedures—numbered steps that guide someone through a process. If you've followed a recipe or assembled a bookshelf or child's toy, you've worked with procedures.

But writing procedures is very different from reading them. Before you write "Step 1," you have to know what Step 1 needs to be. So, before you begin writing the procedures themselves, take one or two steps back and take another look at the process you're describing.

Every process has two points: a *trigger* and an end point. The trigger is what starts the ball rolling, so to speak. It might be a step itself, or it could simply be a condition or state that has been reached. For example, a procedure for how to program the speed dialer in a cell phone won't begin with "Purchase a cell phone." This condition can safely be assumed. The procedure might instead begin with "Make sure your cell phone is turned on."

The end point is the state or condition that happens when the procedure reaches either its natural conclusion or a point at which another process takes over. In the cell phone example, perhaps the last step in programming the speed dialer is pressing the Save button. There are no steps after that simply because there is nothing else to be done. That's the natural conclusion of the process.

Tech Talk

Trigger refers to the action or state of being that initiates a process. The trigger can be included in the first step of a procedure, or it can appear in the paragraph that introduces the procedure. For example, an introductory paragraph might say "Now that you have installed the software, you are ready to configure it." Step 1 then starts the configuration process. Or, Step 1 might read "After the software is installed, begin configuring it by ...."

For example, if you are setting up a database, there might be two sets of procedures: one for setting up the table structures, another for actually loading the data. The end point for the first procedure is the point when everything is ready for data to be loaded. The end point for the second procedure might be when all the data is loaded into the database.

An end point for one procedure often dovetails into a trigger for another procedure. Keep this in mind as you are analyzing processes and identifying triggers and end points. If one procedure needs to follow another, it is especially important to make sure nothing gets left out between the end point of one procedure and the trigger of the next.

But procedures need not be at the nuts-and-bolts level. Sometimes, when the audience is highly experienced, a procedure needs to list only a series of goals in the correct order—the readers already know how to accomplish the interim goals.

Coffee Break

If you think government writing is unclear, check out the Plain Language Action Network at www.plainlanguage.gov/. The Plain Language Action Network is a government-wide group working to improve communications from the federal government to the public. Its recommendations are very similar to those of AECMA Simplified English. The guidance document on that Web site contains some valuable advice for writing good technical documentation.

Dodging Bullets

Use care in how widely you "cast the net" in defining a process. Many processes are made up of other processes that also might consist of yet smaller processes. Choose a level of complexity that is meaningful—but don't bite off more than you can reasonably chew. When in doubt, *less* is more. Two short procedures generally are preferable to one too-long procedure.

Other procedures take in a much greater scope of action than a single process. For example, an overview procedure for a manufacturing process might be triggered by raw materials arriving and might not reach its end point until finished units are boxed, labeled, and shrink-wrapped on skids for shipping.

Once you've identified the trigger and the end point, you have a framework for writing the procedure itself—the steps that get the reader from one to the other. But wait—is there only one reader? It's important early on to establish how many people or "players" are involved. Does the same person do every step in a process? If there's more than one player, who does what, and when?

It can be tricky to indicate a change of players within a procedure. The best method we've seen to handle this is called *playscript procedures*. The numbered steps in a procedure appear in the right-hand two thirds of the page, and a player's job title appears in the left-hand third, aligned with the steps he or she does and appearing only when the player changes. This allows you to address the process flow from trigger to end point but also to indicate who does what, and when, within the process. (For more about this method, see the book *The New Playscript Procedure,* by Leslie Matthies, listed in Appendix B, "For Your Bookshelf.")

Most often, however, you can assume a procedure will be followed by only one person. In that case, you can focus on the steps themselves and not worry about who's doing what. In either case, when you write the steps you must …

1. **Write in imperative voice.** Instructions tell someone to take action in specific ways. If you need to provide special information with a step, include it as a clearly labeled note.

2. **Address only one action,** or one small group of related actions. A common mistake in writing instructions is to try to cover too much ground with a single step.

3. **Make sure each step is an action large enough to be meaningful.** Another common

mistake is to take the "microscope" approach and break a process down into tiny fragments too small to be significant.

4. **Provide obvious transitions or connections.** The reader must never wonder if a step has been left out or how one step follows another. Repeat "landmarks" such as screen names once in a while to keep the reader oriented.

How many steps are too many? Good question. It depends on the subject and context, of course, but many guidelines have said 7 is a good target and 10 or 12 is the maximum. If you find yourself running over this number, try breaking down your long procedure into groups of shorter procedures.

But sometimes the way a product is designed makes it impossible for you to write the short procedures you'd like to. In this case, do the best you can—for example, sometimes combining smaller steps can help shave numbers off the total count. Remember: The more steps in a process, the harder it seems to the reader and the more likely he or she is to get lost.

There's usually no need to indicate the end of a procedure—ideally, the reader can tell the process has been completed. But if there's any doubt, indicate clearly that the reader has reached the end. This might be something as simple as the phrase "End of Procedure" or "Expected Result" on a line by itself after the last step. If the procedure is one of many, you can add a section titled "Next" to the end of the procedure, telling the user what to do next.

Divided by a Common Language: The International Audience

Coffee Break

There are approximately 2,800 spoken languages and another 8,000 dialects throughout the world. Africa has over 700 languages, and India has 18 official languages and hundreds of minor languages.

English has been adopted as the universal language of the computer industry, but don't let that fool you. Having a working knowledge of English and speaking fluent, idiomatic English are two different things. Even within the sphere of native English speakers there are major obstacles to understanding.

Even in countries where English is commonly spoken as a second language, readers won't necessarily be familiar with all the expressions that native speakers take for granted. (There's one right there: "taking something for granted.")

Aside from writing clearly for non-native English speakers, you also might write documentation that is translated into other languages. You can improve readability for an international audience and write for translation by following some simple—but important—guidelines:

➤ Avoid contractions. It might sound stilted and unnatural to a native's ear without them, but contractions are among the greatest stumbling blocks faced by translators and non-native English speakers. It's far more important to be clear, and avoiding contractions leaves no question about the meaning of a sentence.

➤ Write in the active voice. This shows clear relationships between actions and actors, between the subject and verb in each sentence. For a reader perhaps struggling to identify the structure of the sentence as well as its meaning, active voice is a life-saver—and a patience-saver.

➤ Taking a tip from Simplified English, try to use one word to mean one thing. Limiting the use of each word to a single meaning goes a long way toward reducing opportunities for confusion and increasing the likelihood of understanding. Take the word "right," for example: It is an adjective with multiple meanings: "correct," and "opposite of left" to name two); a math term ("right angle"); a noun with multiple meanings ("in the right" and "right to remain silent," for example); an adverb with multiple meanings ("Go right over there" or "It looks right"); and a verb ("to right a wrong").

Keep courtesy in mind. Non-native English speakers aren't stupid. Don't be tempted to "talk Tarzan" to your audience ("Insert disk; press key"). Instead, put yourself in the reader's place and use correct grammar, clear writing, and avoid colloquialisms: the same things that you would hope to see in a technical document not written in *your* native language. Cultural differences can affect technical writing in unexpected ways. Culture can dictate whether you can use a human figure in an illustration, for example, or what colors might be considered inappropriate if used in a technical document.

Writing clearly is the second most important thing you can do to help your reader, closely following knowing the product. Clear writing boils down to being competent, confident, and considerate. If you can do that, the ideas are bound to come through—clearly.

The Least You Need to Know

➤ Follow the clear writing principles outlined in this chapter to focus your writing.

➤ Apply the KISS principle—Keep It Short and Simple.

➤ Write procedures from the trigger to the end point.

➤ Be courteous—not condescending—in writing for an international audience.

Chapter 20

The Right Tool for the Job

In This Chapter

➤ A look at tech writer's tools

➤ A little about computers

➤ What you need to know about desktop publishing—and why

➤ A sketch of graphics programs

➤ Working with the Web

By now you're getting the idea that a tech writer needs to be a jack-of-all-trades—and a master of some. A few short decades ago, a tech writer had a pencil (often the old-fashioned #2 kind), a pad of paper, and a typewriter, and he (it almost always *was* a he) concerned himself solely with writing. How his words were transferred to the printed page and distributed to customers were mysteries about which he probably lived in blissful ignorance.

Today, instead of merely a pocket to hold a pencil, a tech writer needs a wraparound leather tool belt with a good, sturdy buckle. Not only must you be able to interview for information and write clearly and quickly, you also have to know the ins and outs of the magic of document production (and sometimes even packaging design and production). No more handing off a stack of neatly—or not so neatly—typed pages to someone else and turning to the next task on your list.

In this chapter we'll acquaint you with some of the tools and programs on the market. There are many more out there, and you'll no doubt find others you enjoy using. So strap on that tool belt and get ready to pick and choose.

The Tools

As recently as the 1970s, things were vastly different in a tech writer's world: months or even more than a year might have been allocated to write a single document, and the writer needed no knowledge about typography or the printing production process. When computers came along, programmers became writers and their documents reflected the military context in which computers were developed; often writers used military numbering and a military writing style. That was okay, though, because the only people using computers were comfortable in that environment.

The Many Tasks of a Tech Writer

No more. These days tech writers write for a diverse array of readers, including computer novices, and often have to know how to word-process, design and lay out documents, choose typeface sizes and styles, and often draw their own pictures. And tech writers often work directly with printers, sometimes even embedding printer codes into their finished documents before transferring the files electronically to the printer's production facility.

For documents provided with a product on floppy disks or on a CD, there's the issue of providing the text and graphics to the vendor for transfer to the medium, and often there are issues of packaging to be addressed. Need packaging designed and appropriate copyright and licensing verbiage written to go with it? Just ask the technical writer.

Even though none of this production process was any part of a tech writer's job in the 1970s, it is very much a part of a tech writer's day in the computer industry of the twenty-first century. The pace of the process has picked up, too. Tech writers today often must produce, as if by magic, finished documentation ready to ship with the product in much less time than a writer used to just write a manuscript.

Tools, Wonderful Tools!

But don't despair! With the added responsibilities come tools that make most processes easier than anyone dreamed they could be. For example, software now can do file format conversions in minutes that used to take days or even weeks. Need to change an image at the last minute? Open it in a photo-retouching program, make the change, and shoot the new image off via e-mail or upload it directly to the

Tech Talk

JavaScript is a WWW scripting language developed by Netscape Communications and Sun Microsystems, Inc., that is loosely related to Java. JavaScript, unlike Java, is not a true object-oriented language and is not compiled. You're sure to encounter both of these terms, and it will help you to understand what developers are talking about if you know the differences between them. Even better, learn how to work a little in both.

printer's ftp site. Clearly you'll need to know and use software tools most suited to your tasks and those that will help you produce the best work.

As someone responsible for producing documents for your company, your manager might even ask you to recommend what tools your department should buy. With the dizzying array of available tools, it can be hard to choose exactly which one is best, and who better than you to provide insight about the most demanding or critical aspects of your job?

It pays to have as many "strings to your bow" as possible. Today's technical writer probably will know how to use at least these tools:

➤ A desktop publishing tool such as Adobe FrameMaker

➤ A word processing program such as Microsoft Word

➤ A screen capture/image retouch program such as Jasc PaintShop Pro

➤ Adobe Acrobat for creating PDF files

➤ An online help tool such as RoboHelp or SnagIt

And the tech writer with a little "something extra" also will know:

➤ How to create and design Web pages using HTML tags

➤ What JavaScript is and what it's for

➤ At least one programming language such as C, C++, or *Java*

We want to remind you that there's no substitute for doing your own homework about what's out there, what's used in your geographic area or at your company (or a company you want to work for), and what fits your needs best. New software comes out with staggering frequency and something new and wonderful we haven't covered here could be released tomorrow. We are, after all, talking about the computer industry.

We also must emphasize that we are not recommending one program over another. If we don't

Dodging Bullets

Learning how to use the tools means you should be able to do something useful with them beyond the merest basics. Explore their capabilities, options, and features so even if you're not an expert you still have an idea of what they can do. Don't say you know how to use a program if all you can do is open and close it.

Tech Talk

Java is an object-oriented programming language developed by Sun Microsystems, Inc. Similar to C++ but easier to use, Java is a popular language for writing small applications, called "applets," used in pages on the Web.

Dodging Bullets

Don't make the mistake of saying "computer" when you really mean operating system. An operating system (OS) refers to the main control program of a computer that pretty much does all the computing. Some PC operating systems you might be familiar with are DOS, Windows, NT, UNIX (pronounced *YEW-nix*), and the Apple operating system, called Mac OS.

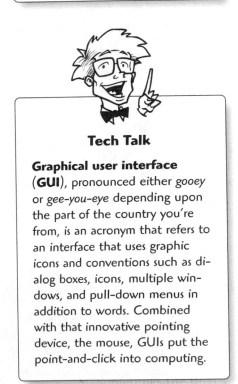

Tech Talk

Graphical user interface (GUI), pronounced either *gooey* or *gee-you-eye* depending upon the part of the country you're from, is an acronym that refers to an interface that uses graphic icons and conventions such as dialog boxes, icons, multiple windows, and pull-down menus in addition to words. Combined with that innovative pointing device, the mouse, GUIs put the point-and-click into computing.

mention a tool here, it's not because it's bad—it's just that we have limited space and are presenting examples of the types of tools you'll want to research. Our purpose is to attempt to describe some of the tools that tech writers around the world are using today.

Your Computer Is Your Best Friend

After the tool in your head—your brain—the next most important tool for your job is your computer. We assume you're interested in or at least comfortable with computers, or you wouldn't be reading this book. But you might still be wondering which type of computer to buy (for those inevitable weekends you'll need to work at home), and which operating system to learn. There are a lot of choices, and you might be afraid to make the wrong one.

For the office, your employer will provide a computer as well as whatever software tools you should use. You need to know how to *use* those tools when you walk in the door.

The good news is that if you can use one computer system, you can easily learn how to use another. Most, though not all, of the basic premises are the same, although the user interface might be different. The biggest difference is between Windows-based operating systems rooted in MS-DOS (Windows 95/98/Millennium/2000/NT) and those rooted in UNIX (Solaris and Linux).

Graphical user interfaces have done a lot to make the user experience fairly standard from one system to another. There are still important differences, though: The Linux desktop might look a lot like a Windows desktop, but start pulling down menus and looking for familiar Windows taskbars, and you won't find them.

The most radical difference in user experience comes when (or if) you start working on the command line, typing specific commands next to a "prompt" on an otherwise blank screen. Working on a command line is strictly for experts, mostly because you can't do anything unless you know what commands to type and how to control and combine them to do what you want.

You probably can do fine as a tech writer if you're familiar with your PC and only some flavor of the Windows operating system. In our experience, the majority of writers in documentation departments use tools that run on a Windows OS (Windows NT, Windows 98, and Windows 2000 are common).

Actually, it's seldom up to you to choose what computer or operating system you'll use—your employer makes that decision for you, often based on the operating system the company's product runs on or is compatible with. For example, although UNIX-based computers are not as common in documentation departments as Windows-based, they are still heavily favored at companies such as Sun Microsystems and AT&T, which develop UNIX-based products.

The Value of Being Familiar with UNIX

Speaking of UNIX, tech writers who want to succeed should learn something about it and how to use it. Why? Mostly because historically the most common operating system behind network servers has been UNIX, and that's also true of servers attached to the Web. Programming that serves Web sites, such as e-commerce software, shopping carts, and user-customization products, have necessarily strong ties to UNIX.

... and Linux

Beyond that, knowing about UNIX gives the writer knowledge of *Linux,* which is an open-sourced form of UNIX that took off like wildfire in the late 1990s. Its ready availability (versus the high cost of UNIX) essentially puts the power of UNIX into the hands of anyone who cares to install it and learn it. Companies are scrambling to build applications for it, and there are bound to be a wealth of opportunities available for Linux-savvy tech writers in the next 5 to 10 years.

Tech Talk

Linux is an open–source implementation of UNIX created by Finn Linus Torvalds, which runs on many different hardware platforms. Assorted, completely legitimate forms of Linux can be downloaded free from the Internet, as well as purchased as part of a distribution on a CD-ROM. A popular Linux version is Red Hat Linux.

Leave the Apples at Home

If you're thinking you can get a leg up by learning to use your mom's Macintosh on weekends, think again. With a few exceptions (for example, companies making applications that run on Mac OS), Apple knowledge is not in high demand in the tech writing world—a world increasingly focused on Internet-driven or Internet-delivered products and services running on systems powered by Windows NT or UNIX. Sorry, Mom.

No matter which operating system you use, the bottom line is that you are comfortable enough with it and that you can confidently run basic applications you'll use every day for functions such as word processing, handling e-mail, and browsing the Internet.

Writing Tools Must Fit Your Project

If you're just going to the grocery store and back, you don't need a Ferrari. Likewise, a Ferrari is not a good choice if you're transporting a young soccer team and all their equipment. You only need a Ferrari when you need what a Ferrari is good for—going fast, hugging curves, and, well, showing off. The same is true about the tools you'll use for projects at work.

Tech writers argue a lot about whether an employer has any right to require expertise in any particular publishing tool. A good writer, they say, is hired to write, and is smart enough to learn any tool to do that job. After all, driving is driving, right?

It's true that an Interleaf user can learn FrameMaker, or that a Word user can learn QuarkXPress. But employers have reasons for looking for specific tool knowledge. First, it's not as simple as handing you the keys and saying, "Go!" Getting the most out of a dump truck requires different skills from those that get the most out of a Ferrari, and vice versa. There's always a learning curve, and time, as we said earlier, is the thing in shortest supply in the computer industry. Being willing to learn is not enough. You need to already know.

Another reason is one of sheer practicality: The people in the personnel department are expected to screen, interview, and hire for all positions (writers among them), and they usually don't know in an abstract way what tech writers should be able to do. It's much easier for them to go by a list of desired tools provided by the manager the writer will work for.

Desktop Publishing

The characterizing feature of a desktop publishing tool (DPT) is that it allows you to use a computer to create camera-ready copy for printing. A DPT is considered to be different from a word processor because word processing is only the text-preparation part. These days, however, word processors (such as Microsoft Word) incorporate a lot of the features previously designated as desktop publishing features, and the lines between these two types of tools have become thoroughly blurred.

But they aren't obliterated entirely. With DPT you can design page layouts, import text and images, and often create books from multiple chapters. Word processors still focus mainly on creating and manipulating text, with other capabilities more or less tacked on. But just as video gamesters have their favorite joysticks, tech writers have their favorite tools—even though, ultimately, you can produce documentation of some sort with any tool.

Another twist on the DPT/word processor debate is geography. The demand for desktop publishing differs somewhat around the country. For example, if you work on the West Coast in the United States, you are likely to be expected to use FrameMaker; if you work in the Midwest or the South in the United States, Word might be the choice of the local employers. So, for you, the question also becomes one of what tool employers in your area are using.

The best response to the FrameMaker/Word dilemma is to simply learn both. If you already know one, working expertise with the other should come quickly. The fact that you've already eliminated that learning curve for either product will make you that much more attractive to prospective employers. And if there's a third program an employer wants you to use—for example, Interleaf—you've already shown you can teach yourself. The following list of tools gives you our take on what a writer needs to know about what's happening in the DPT world. As always, something new might be introduced tomorrow, but we don't think any of these will be going away soon.

Here's a brief summary of some of the most common DPT tools:

➤ **Adobe FrameMaker.** One of FrameMaker's great strengths is how robust it is in creating large books from multiple files. Another is that it runs on Windows, NT, UNIX, and Macintosh, and files created on one platform can be used on another. FrameMaker 6.0 introduced a feature that lets you apply document commands easily across an entire book. It also comes bundled with Quadralay Web-Works Publisher for quick document conversion into HTML or XML. Is FrameMaker easy? No. Intuitive? No. But writers who use it are it's very strong advocates, and FrameMaker moved up to the head of the DPT class very quickly.

➤ **FrameMaker + SGML.** This variation on the standard FrameMaker version lets you create documents in SGML (Standard Generalized

Pros Know

"Technical writers who are familiar with a variety of software tools like desktop publishing, online help, graphics, presentation, and Web authoring software, greatly increase their marketability across a broad spectrum of potential employers. Their exposure to many different software products—and the fact that learning to use the software casts them temporarily in the role of end users of documentation written for them by others—also gives them valuable insights into product and document usability."

—Frank LaCombe, President, KnowledgeTree Systems, Inc., www.k-tree.com

Markup Language), a structured format that can be converted to XML and HTML as well. If you need to use SGML in your business, this FrameMaker version is a relatively easy-to-use interface; it also lets you use standard FrameMaker. But the two versions aren't interchangeable—conversion from FrameMaker to FrameMaker + SGML can take some work.

➤ **Adobe PageMaker.** Another Adobe product, PageMaker is good for small documents that need more advanced layout and graphics features than FrameMaker offers (brochures or marketing pieces rather than large, text-intensive books). It runs on Mac OS and on standard flavors of Windows.

➤ **QuarkXPress.** QuarkXPress has been used in documentation departments, but we see it used less for technical documentation these days. It is still in demand, however, for advertising, magazine, and catalog work because it's great at handling graphics and layout. Some users feel it's limited in its ability to handle large document files. Quark runs on Mac OS.

➤ **Corel Ventura.** This DPT has some very powerful layout features, although its book capabilities are not as strong as you might need if you are working on a large document set. Ventura comes packaged with WordPerfect, Corel PHOTO-PAINT, and other programs, giving you full DPT capabilities in one box. Corel Ventura runs on Windows operating systems.

➤ **Interleaf.** Although the Interleaf company has been bought by BroadVision and its publishing software no longer is being sold, many companies still use it—Interleaf required a very large financial investment and it's a powerful tool that takes some time and work to convert to another format. It's a complex program, but it has excellent book-building capabilities for managing large documents and document sets. It runs on UNIX and Windows operating systems.

Pros Know

When choosing a DPT for typical documentation, layout capabilities are not the number-one thing to look for. Technical documents need to be uncluttered by fancy layouts, "eye-candy" graphics, or attention-getting type styles.

Different companies might use programs other than those we've listed here, but these are what a potential employer might reasonably expect you to know. Don't feel inadequate if you don't know how to use a tool you've never heard of.

Word Processing Programs

Many, many companies use a word processing program instead of a DPT to create their technical documentation. Microsoft Word is good and seems to be used almost universally; Corel WordPerfect also is good, although we don't hear of it being used much in technical publishing.

Although the primary function of these programs is to create and manipulate text, they nevertheless do much more than simply let you type: You can create layouts, import images and graphics, and convert native documents to other file formats. Still, their main strength is in managing text, so bear this in mind when you use them. A good word processing program should provide basics such as a spell-checker, grammar-checker, thesaurus, autocorrection capability, and of course WYSIWYG (What You See Is What You Get) display of pages both as you type and as a preview of the printed page (print preview). Microsoft Word is pretty much the word processor benchmark to use for comparison.

One caution we feel duty-bound to offer, however, is that Word has not so far proven it can handle multiple-file books larger than about 200 pages gracefully. We hope this will change in some future release. If your company needs to produce large documents assembled from many files, this limitation would be sufficient reason to choose or recommend FrameMaker over Word, if you have a choice.

Choosing the Program That's Right for You

The pragmatic truth of the matter is that the program that's "right" for you usually is the one your company has already bought. If your employer is happy with a particular tool and the documents are being successfully produced, your best strategy is to learn that tool (if you don't know it already), and learn it well.

Once you become a "power user" of any publishing software, however, you'll be in a good position to research alternatives and learn whether they might be preferable. When recommending a change in software to management, though, think carefully about *all* ramifications. File conversions in particular are an important consideration; they can introduce serious errors or make tedious (meaning costly) hand-correction necessary. You might be itching to use another tool, but the cost of the new software, combined with the cost in time and effort to *accurately* convert existing files, might be more than your company is willing to pay.

Pros Know

Human beings are by nature highly visual. Many SMEs you'll work with think in images rather than words. Bear this in mind when an engineer starts drawing—instead of talking—in response to a question. Accurately copy the drawing into your notes: it holds the key to the SME's answer.

Graphics Tools Illustrate Your Points

Today's tech writer also needs to have some graphics skills in his or her bag of tricks. "Graphics" is an elastic term that encompasses a lot of things—drawing freehand, developing charts and graphs, taking and editing screen captures, designing layouts, and perhaps even photo retouching or image manipulation.

A good visual can increase the information value of a document tremendously. How? The combination of a graphic, high-level overview of a concept or process, and information-dense text that spells out the details gives readers multiple levels of understanding on the same page or pages. Illustrations also make it possible to show relationships that would be difficult or lengthy to explain, packing more information into a smaller space.

Illustrations also appeal to a less technical audience. Readers who might struggle with an unfamiliar vocabulary can quickly grasp a straightforward illustration. This understanding then makes the information in the text more meaningful and accessible.

A diagram also can act as a quick reference for a long, complicated procedure. The following figure shows a flowchart of just such a procedure, for which the explanatory text is nine pages long. The illustration summarizes the process on a single page—a user can scan the diagram for a quick understanding or refer to it after reading the procedure.

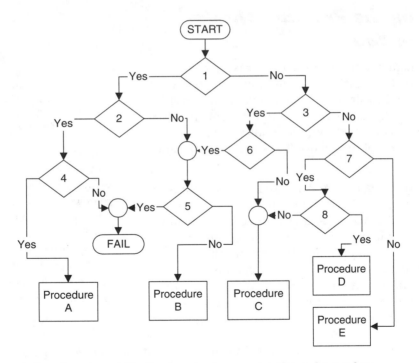

A flowchart is an effective way to show a complicated procedure or process.

What is a screen shot? Nothing less than one of the most common—and useful—forms of illustration in any user manual. A screen shot (or screen capture) is a picture of the window or dialog box from the program being described. As the user works through the steps in any task, he or she can compare what's actually onscreen with what is supposed to be there. The user knows he or she is on track and that the process is unfolding as it needs to.

Chapter 21, "Document Design," contains more information about using illustrations and screen shots effectively in a document layout.

Image Files

Whenever you create a graphic, you must save it in a file format that's appropriate for your document. For example, documents destined for paper need a graphics format that the printer's equipment can read, but documents meant to be read online and printed in fragments at will by the user need figures in a format compatible with this kind of use.

There are two types of illustration formats you need to know about: raster (bitmap) and vector (line drawings). You'll create raster graphics with programs such as PaintShop Pro and Photoshop, and vector drawings with programs such as Adobe Illustrator or Visio.

Importing a graphic can mean either placing the contents of the graphics file directly into the document file (copying), or placing a pointer into the document that links to the graphics file ("importing by reference"). The choice of which method to use is not trivial—image files can be huge, and copying them directly can make documents balloon to unmanageable sizes. Graphics are harder to manage when they're copied directly, too. Importing by reference allows you to change a graphic at its source, and the changes are reflected in each file where it appears.

This isn't a definitive list of file extensions for all possible graphics file formats, but these are some of the formats you need to be familiar with. We list these because they are usable in just about every image conversion program, and even the Web-based formats can be imported into all leading word processors and DPTs:

➤ .BMP (BitMaP). High-resolution image file used for print documents.

➤ .TIFF or .TIF (Tagged Image File Format). High-resolution image file used for print documents.

➤ .GIF (Graphics Interchange Format). Image file used for the Web (the advantage is that it lets you create transparent backgrounds for interesting graphic effects).

➤ .JPEG or .JPG (Joint Photographic Experts Group—a standard for storing images in compressed form). Image file used for the Web, with more flexibility for creating higher-quality (and bigger) image files.

Your company might use only one type of image format in its documents for consistency. If it isn't obvious what format or formats you should use, ask your manager. If there's no preference, we recommend using .TIFF for images in your printed documents. For Web documents you must use .JPEG/.JPG or .GIF files, because these are the only formats currently supported.

Screen Captures

To create screen captures, or screen shots, you need to use a program that takes a picture of the computer screen and, if necessary, crops it and converts it into the proper format. There are many such programs, a lot of them available free of charge or at a reasonable cost on the Internet. PaintShop Pro, SnagIt, Visual Capture, and FullShot are just a few.

A good basic screen capture utility will let you

➤ Save the image into a variety of file formats, including .TIF, .GIF, and .JPG.

➤ Crop the image (cut off parts, leaving only the area you want).

➤ Include a delay feature so you have time to open a menu that needs to be included in the shot.

It's also nice if the utility lets you do simple retouching and painting, but those are extras and are functions you usually can perform better with a different program.

Pros Know

If you don't have a special screen-capture program, you can use the built-in screen capture utility available on your PC desktop. To capture an image of your entire computer screen, press the **Print Screen** key. To capture an image of only the currently active window, hold down the **Alt** key and press **Print Screen.** Either of these methods copies an image into the "clipboard" on your desktop. To paste the image into your document, just open your document and choose **Paste.**

Photo-Retouching Programs

There will be times when an image is less than picture perfect, and for those times you need a photo retouching program such as Adobe PhotoShop or Corel Photo-PAINT. These types of programs allow you to do things such as change colors, add text, improve the quality and sharpness of the image, reverse all or part of the image, even merge multiple images together.

Serious image manipulation, like drawing, is a skill that requires a lot of practice and learning—probably more than you'll have time or inclination for. But it's still worthwhile to learn a few helpful tricks. For example, maybe there's a screen shot that shows the name of an engineer in your company, or that needs different data displayed—you'll need to know how to "paint" something out or drop in new information.

Or maybe your document is ready to go to print when you learn the developers have revamped the user interface. Instead of taking all new screen shots and importing them into the document, you have the option of simply modifying the existing images like in the example shown here (whew!).

Enter the path for your home directory:

C:\Home

Browse...

☑ Allow anonymous access to this web site.

< Back | Next > | Cancel

You create a screen shot and put it in the document. Right near release time, when you no longer have access to the software and no time to make new screen captures anyway, you learn the buttons have changed and the example is incorrect in the text field.

< Back | Next > | Cancel

Enter the path for your home directory:

C:\ProgramFiles\Ebuyandsell\cgi-bin

Browse...

☑ Allow anonymous access to this web site.

Here is the second screen shot, made from the same file by retouching the image. The buttons and margins were moved and a different path name appears in the text field. No need to open the software and create a new screen shot.

Image retouching is especially useful when your screen shots include a developer's name or a real person's e-mail address.

You'll be glad to have an image-retouching program when the user interface changes at the last minute. These two dialog boxes were created from the same image file.

Drawing Programs

What used to be done with pencil or pen and ink on drafting boards is now done almost exclusively with drawing programs. If you have an artistic bent, these tools can be a lot of fun to use, even if you don't create the original drawings yourself (which you seldom will). Technical illustration at a professional level requires a lot of knowledge and a special kind of talent; creating technical drawings, such as for printed circuit boards or hardware parts, also can take an enormous amount of time, even for skilled illustrators.

It's not a mistake to add a good drawing program such as Adobe Illustrator or Corel Draw to your tool belt, but don't expect yourself to master it quickly. And you might never develop the expertise that a technical illustrator has. For this reason, it's far more cost effective to hire professional illustrators to create any but the simplest drawings for your documents. You can give a competent artist a pencil sketch or a simple prototype diagram you've done with your tools and receive back a professionally executed drawing in the shortest time frame possible. If you're really good with your drawing tool, you might be able to make any subsequent changes to the drawing yourself once it's been created for you.

In fact, using a drawing program to make corrections or changes to existing drawings probably will be your primary use for this tool. For example, when updating a document for a new release, an illustration might require only a couple of new callout boxes or arrows, or a new button or menu option—you should be able to make these types of changes.

Just between us, we think Visio is one drawing program worth getting under any circumstances. Visio includes all the elements you need to create diagrams, charts, schematics, and every type of technical diagram. It's an exceptionally handy tool at a modest price.

Online Tools Launch You into Cyberspace

With the trend moving away from printed documents and toward CDs and embedded help files, and with the Internet and its kindred technologies seeping more and more deeply into our society, you'll probably find yourself—sooner rather than later—developing some kind of documentation to be delivered online. Chapter 22, "WWWriting for the Web," discusses writing for the Web, but before you can do that, you need some tools. You've come to the right place.

Creating PDFs

A growing practice in the computer industry is to make documentation available only in the form of .PDF (Portable Document Format) files. PDF is a universally readable file format developed by Adobe Systems. It captures formatting information so that a document appears onscreen or on a printer exactly as it appears in your original. A user does need Adobe Acrobat Reader to view .PDF files, but Adobe Systems has very sensibly made Acrobat Reader free and available for download on the Web.

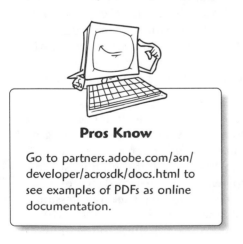

Pros Know

Go to partners.adobe.com/asn/ developer/acrosdk/docs.html to see examples of PDFs as online documentation.

You can convert any document to a .PDF file (also called simply "PDFs," which we'll do from here on) using Adobe Acrobat software (Distiller or PDFWriter). Because PDFs are meant to be viewed or printed as-is, you can do little or no editing once they're created (whether you created them or someone else did).

It's worth noting here that PDFs seem to have single-handedly changed the direction of online documentation. A few years ago when "the end of the printed manual" was predicted, many companies actually planned to put all their documentation online and assumed it would be used and read there as well. The idea was to convert it all into HTML or SGML, and technical publishing departments got busy planning the structure of this new form of documentation.

But then PDFs came along and seemed to do away with those other plans. Why? Most importantly, they are easy to produce, having been created directly from the same files used to create your printed documents. And because even when documents are intended to be used and read online, companies discovered that people invariably want to print them (or parts of them) as well.

PDFs introduced an elegant way to provide documents that looked beautiful online and that printed cleanly as well. PDFs were the vanguard of *single-source publishing,* allowing users to choose between printed and online documentation with a simple click of a button.

If you create your original document in FrameMaker, Word, or any other program that converts easily into a PDF, all the cross-references and links in the original document are preserved in the PDF. The user can click an active link to jump from the page number in a table of contents to the page itself, for example, or from an index entry or a cross-reference to the actual item referenced.

Tech Talk

Single-source publishing refers to the ability to create all your user documentation, specifically printed and online, from a single original file.

HTML Editors

Things have changed since the days when anyone who wanted to create a Web page had to learn to use HTML tags—manually placing beginning and ending markers, or "tags" around text. Now there are a number of Web page composers on the market, notably Microsoft FrontPage and Macromedia Dreamweaver. These tools provide overall Web site management tools in addition to WSYIWYG HTML editors, but you can still use the editor even if you don't need the other capabilities. Version 5.5 of Internet Explorer comes with just the editor from FrontPage, and Netscape Communicator comes with an easy-to-use HTML application called Composer, so the tool you need might already be on your desktop. You still need to know what the HTML tags are and which one does what, but for a newcomer to HTML, it's a vast improvement over typing every character yourself.

For full sets of Web documentation, you might consider a program such as Webworks Publisher Professional. Bundled with FrameMaker 6.0, Webworks lets you convert FrameMaker documents into HTML.

So do you still need to learn to code HTML by hand? Yes and no. Yes, because it empowers you to tweak and massage generated pages for an effect you want or to solve a problem. Most important, it lets you understand what's going on in the source files of an HTML page created by the developers at your company. No, because with so many quick and easy tools, you could create Web pages for years without ever having to look at an actual HTML tag. It's a good skill to have, but think of it as one of those "something extras" rather than a foundation skill.

Web Development

Developing pages for the Web demands much more these days than just generating static pages with text and images—pages need to do more, and that means scripting. As we mentioned at the start of this chapter, the tech writer with the extra "oomph" will know something about JavaScript, a scripting language loosely related to Java that's designed to work with the World Wide Web. It brings Java capabilities to your HTML pages—the power behind common page components such as animations and banner ads.

Piggybacking on the popularity of Java, JavaScript has become one of the most popular languages programmers use to create Web pages for Internet businesses. Your knowledge of JavaScript will help you in reading their HTML files, in designing and creating Web pages yourself, and maybe even in getting the job you want.

SGML (Standard Generalized Markup Language)

HTML, XML, SGML—is there a pattern here? The letters "ML" that appear in all of these stand for "Markup Language." The way a markup language works is by using paired tags to say something about (or "mark up") the text between them. A pair of tags might mean "the text in here is a heading" or "the text in here is a paragraph" or even "the text in here points to a hyperlink on the Internet."

The X or HT or SG tells you what "flavor" the markup language is. SGML is an internationally agreed-upon standard for information representation that focuses mainly on structure, and is a meta-language (language that defines other languages). SGML is the "mother of all markup languages" in that HTML is a subset of it and XML is a "knock off" of it, sort of a newer "me, too" meta-language (see the "XML [eXtensible Markup Language]" section later in this chapter).

Files called DTDs (Document Type Definitions) accompany SGML files. They contain all the information needed to accurately render the entire document, down to specifics like font type and size. For example, a DTD for customer documentation might specify that all files must start with a component called frontmatter, then have a Heading 1 followed by an introduction, then a Heading 2, and use Times Roman 11 point as a typeface.

The DTD for online documentation, by comparison, might specify a different structure and a different typeface. There could be yet another DTD for marketing documentation, and still others for different types of documents.

The value of this is that by applying a different DTD to the same SGML file, a document is instantly reformatted. If later your company, say, changes its page design, a different DTD can be used to apply the new look and structure to existing SGML pages without retyping a single character.

XML (eXtensible Markup Language)

XML (eXtensible Markup Language) is designed especially for the Web, with the ease of HTML and the structural advantage of SGML. XML is rapidly emerging as the universal format for structured documents and data on the Web. For that reason, you're quite likely to encounter it in your tech writing travels.

XML, like SGML, is a meta-language that defines markup languages to create structured documents. More flexible than either HTML or SGML, it enables you to create customized tags that add meaning to the components wtihin the tags. You get to define not only the document's look and feel, but also some of the functionality.

Help Tools Mean Really Ready Reference

Another solid tool for your toolbox is an online help tool. We discuss online help in Chapter 22. Three of the major help tools in use now are Doc-to-Help, ForeHelp, and RoboHelp. You can't go wrong learning any of these.

Wrapping It Up

Whew! Bet that tool belt is feeling heavy by now. As you can see, there's a lot more a tech writer can be expected to know than how to write—and although we've covered the big topics, we've barely scratched the surface of available tools. Appendixes B, "For Your Bookshelf," and C, "Professional Organizations and Web Sites," of this book include, information for books and Web sites to get you started.

And remember to use your own investigative skills to discover and check out other tools we haven't mentioned here. That's the tool you'll use most, to help you choose and use the others wisely.

The Least You Need to Know

➤ Think outside the writing box—you need far more than writing skills to be a successful technical writer.

➤ Think Windows: The kind of computer you'll use depends on what company you work for, but it's a safe bet you'll need to know how to operate a Windows operating system.

➤ Equip yourself with the basics: a DPT (such as FrameMaker), a word processor (such as Word), a screen-capture utility (such as FullShot), and Adobe Acrobat.

Chapter 21

Document Design

In This Chapter

➤ Adding good design to your tool belt

➤ Producing documents that are easy to use

➤ How to find inspiration from someone else's ideas

➤ What a grid is and how to create one

➤ Putting it all together

You have no training in graphics, or design, or art-anything. When you were in grade school, the teacher took your crayons away. But now you're a tech writer and your boss just "asked" you to come up with a clean, solid layout for the new manual!

"Hey, I'm a writer, not an artist!" you might say. We have news for you: Writing isn't all there is to being a technical writer. Don't panic—you can do this (yes, you). Just like writing, designing is based on known, learnable guidelines and principles. Of course, we can't teach you everything there is to know about page design, layout, and typography in a single chapter. But we can equip you to proceed with confidence in creating visually effective printed documents, PDFs, and online documentation. Nothing is as hard as you think once you know how it's done.

Whose Job Is It, Anyway?

Desktop publishing systems have made it possible to create professional-looking documents of all kinds cheaply and quickly. A technical publications department that once had a full array of illustrators, layout artists, and typesetters might now have only a single tech writer who works with FrameMaker and Photoshop. The downside is that this often means the tech writer has to be the artist, too. This usually means taking the full weight of responsibility for creating the design and layout of whatever documents the department produces. So whose job is it? Yours.

If the technical writer happens to also be a graphic artist with book design experience, this is fine. Is this you? We didn't think so. It's far more common for the technical writer to have no idea how to design a book or an online page. Needless to say, this can be an exercise in frustration for the writer and, ultimately, result in a final document that falls far short in appearance.

Get Help When You Can

Illustrations can greatly enhance a document by illustrating concepts that are difficult to describe. But all illustrations are not alike. Although you definitely can learn the basics of layout, typography, template creation, and simple graphics skills, technical illustration is a whole different animal. Before we tell you what you *can* do yourself, we want to make sure you understand what you *can't* do, and shouldn't even attempt.

Sure, you can do some charts and drawings with the drawing tools on your publishing software, or with your copy of Visio. There's plenty of copyright-free clip art floating around these days, too. But complicated illustrations that clarify your text, the ones with lots of components and complex relationships, need a level of visual talent and experience that seldom are found in a tech writer's bag of tricks. You literally can spend days laboring on a drawing that might take a skilled technical illustrator only minutes or a few hours to create. It's not only impressive to watch, it's a lesson.

Enhance Document Usability with Visual Elements

As we mentioned in Chapter 5, "What Makes a Good Document?"; "usability" is a word you'll hear a lot during your tech writing career. There are two areas of usability that concern the tech writer: the usability of the product itself (and as you write about it, you help to make it more usable), and the usability of the documentation about the product. You have, of course, more control over the usability of the document.

Visual effectiveness is an important factor in usability. A poor design can create confusion when instructions are buried in blocks of text, warnings and cautions are hard to identify, and text is just too small. A document that is too difficult to read won't be read—and even if it is read, it will not be properly understood.

A good design can turn a bad document into a good one, with exactly the same content. How? Adding white space makes reading easier because the eyes aren't overwhelmed. Clearly marked headings designate structure within the document. Clearly marked procedures are easy to find and to follow. Well-done illustrations depict an idea far more succinctly than text does, conveying essential information clearly and quickly.

All you need to create visually effective documents are your desktop publishing tool or word processing software, image manipulation software (all discussed in Chapter 20, "The Right Tool for the Job"), and a little design know-how. This chapter is about this last item.

It's easy to begin. Your design should work toward a single goal: ease of use. If your choices are between "boring but clear" and "splashy and exciting, but hard to read," it's a given that "boring but clear" wins hands-down every time. Exciting and innovative graphic design has its place but, unfortunately, it isn't in technical documents.

Pros Know

Design is a learnable process that follows known guidelines and principles; four of the basic principles of design are ...

➤ **Location** (place related things near each other).

➤ **Alignment** (create visual connections between page elements).

➤ **Repetition** (repeat visual elements such as fonts, icons, and rules).

➤ **Difference** (use visual contrast to indicate contrasting meanings).

But that's not because technical documents are used by people who don't appreciate color, jazz, and eye appeal. The reason is far simpler than that: To convey information clearly, a document should not contain any distractions nor any elements that make it harder to read. Nearly always, the purpose behind any gee-whiz visual is to successfully compete with other things vying for a busy reader's attention. Technical documents don't face that challenge—our readers aren't going to turn the page to the next ad, or pick up a different brochure from a sales counter, for example.

An easy-to-use document has the following characteristics:

➤ A size that's easy to handle—neither too large nor too small

➤ A binding that lets it lie flat when opened and that keeps the pages from falling out

➤ A cover that indicates clearly what product it goes with, what company made it, and what type of document it is (tutorial, operations reference, and so on)

➤ Easy ways to find the information you need, through means such as a clear table of contents, tabs, and headers and footers

➤ A thorough index (as opposed to an index that's an afterthought, or reads like one)

➤ Typefaces that are clear and readable

➤ Visual elements such as graphics, illustrations, tables, and notes that are meaningful and used thoughtfully and consistently

Dodging Bullets

When you start working with an idea for how a page should look, don't feel as if you have to fill up every square inch. The space you leave around visual elements actually conveys information by separating items that are not as closely related. It also eases demands on the eye, and gives the mind fewer things to take in, assess, and assign meaning to.

Coffee Break

The Society for Technical Communication (STC) has competitions for technical art as well as technical writing. Art categories include illustrations, photographs, document designs, posters, and packaging design.

See How Others Have Done It

The best advice we can give you in developing a document design is to borrow from the experts. We're not talking about stealing, but learning. Look through several technical manuals, stopping to think about what makes one a pleasure to read whereas another is frustrating or confusing. When you find a layout you like, analyze the elements and try to create samples of your own company's content that have that same ease of use.

We don't, however, endorse or recommend duplicating another company's look exactly—first, you don't want your company's documentation to look like someone else's, and second, you don't want to raise issues of copyright violation. But you certainly can borrow ideas such as type size, column width, and heading size from various examples.

If you've been to art school, you already know one big artists' secret: A lot of design is simply trial and error. The trick is recognizing when something you've tried succeeds.

Page Setup

Page setup determines your basic document size as well as the image area, the area on the page or screen where you place your print, and graphics. Every page of a document needs to indicate clearly the title of the document, the chapter number and name, and the page number. Typically, a chapter name goes in the upper outside corner and page numbers appear at the bottom outside corner. There are many other aspects to take into account when doing page setup, which we discuss in the following sections.

Page Size

Do you produce paper documents or electronic only? If you produce paper documents, who prints them? And how? If printing is done on a photocopying machine, your page size obviously will be something that the photocopier can manage, such as North American letter-sized (8^1/$_2$" × 11") or A4 (21 × 29.7 cm), depending upon where you live. If your documents are sent to a printing house, you have more options. Consider creating a smaller page size. A smaller size such as 7" × 9" or A5 (14.8 × 21 cm) is easy for a user to handle, and with the right margins, works when photocopied on any international paper size.

Think about how much and what kinds of information you need to fit on a page. If your documentation is full of wide tables and huge illustrations, you might need a large page size. If the content is mostly text, you can use a smaller size.

If your documents are distributed to a worldwide audience, whether on paper or as PDF files, you can bet people also will be printing them all over the world. If your page size is North American letter size, make sure you create a layout that looks good when printed on A4 paper as well. Stick with the standard "portrait" orientation for a page. People who want to print from your PDFs might not be able (or know how) to adjust their printer settings to accommodate a nonstandard print orientation.

Margins

If your page is large, make your *margins* and *gutters* fairly wide, but this is a balancing act. Nice, wide gutters and margins make a page easier on the eye, but they can eat into the page space needed for content. Pages with too little content can be just as tiring and bothersome for readers as pages with too much. If you find yourself struggling for a happy medium, choose on the side of less information than more.

The choice of bindings also affects the column width and margins. For example, if the document will be in a three-ring binder, the gutters need to be wide enough that the punched holes don't interfere with the text. We've seen plenty of great illustrations marred by a binding hole drilled right through a key element or callout label.

Tech Talk

A **margin** is the white space on the outside of the image area (top, bottom, and outside on a bound page, or all four sides on an unbound page).

A **gutter** is the white space on the edge of the page toward the binding. This term also refers to the space between columns.

Tech Talk

Stock is the term printers use when talking about any kind of paper, from the lightweight paper you use in your laser printer to heavy, coated paper used for covers.

Pros Know

"Before a project even begins, spend some time brainstorming with your print partner. Have the printer do some sample file transmissions and print runs to see if the capabilities of the printer meet your expectations. The most important thing you can do to guarantee success for your technical documents is to create high-resolution PDF files. This greatly reduces file errors when submitting a job to the printer."

—John Mahoney, Mahoney Printing Services, www.mahoneyprint.com

Bindings

Like page designs, the bindings you use for technical documentation might not be flashy but they are as functional and reliable as a pickup truck and won't let you, or your user, down. Discuss your binding options with your printer, or with the person who handles production at your company.

➤ **Perfect binding:** The pages are single sheets glued or pasted together at the book's spine. This is good for books with fairly high page counts; books with fewer pages might have trouble staying open, or can soon come apart when users flatten them to overcome their tendency to close. This permanent binding method lets you print titles on the spine.

➤ **Comb binding:** A plastic "comb" holds single pages together, usually down the left side but sometimes across the top. This is not as sturdy or permanent as the other binding methods, but does have the advantage that titles can be printed on the comb. Also sometimes called "GBC" binding, this method is less expensive than many other methods and also allows the book to lay flat.

➤ **Double-loop wire binding:** A sturdy, permanent spiral wire, often coated with plastic, binds the pages and covers together, usually down the left edge but sometimes across the top. This binding system creates documents that lie flat at almost any page count. Pages can tear out more easily than with other methods, however, so a heavier paper stock might be called for.

➤ **Saddle-stitch:** Sheets printed as double-page spreads are folded in half and stapled at the fold to bind the document. This is best for documents having no more than 32 pages.

➤ **Three-ring binder:** This old standby consists of covers attached to a spine that has sturdy, spring-loaded rings to hold a stack of loose sheets. Each sheet has holes punched (or "drilled") in it to match the size and spacing of the rings. Binders are great if you need to send out regular update pages. The user can insert the changed pages and get rid of the obsolete ones, and you don't have to reissue an entire document. The downside is that you have to rely on the user to actually do this. Binders can be imprinted on the front and spine, or you can get binders with clear pockets to hold colorful cover and spine inserts.

Dodging Bullets

Think about the users when choosing the binding method for your printed documents. If users typically use the document in one place, such as at the office, something less durable might be acceptable. If, on the other hand, the users might carry it around in a laptop bag or briefcase, make it capable of standing up to this harder wear. It might cost more initially, but in the long run a more durable binding pays for itself in user satisfaction.

Page Layout and Design

Once you've decided on the page size, margins, and binding method, it's time to create the page layout. A layout is the plan for how text, headings, and graphics are to be arranged on a page or in a visual space.

Use a grid to help you with your design and to help maintain consistent placement of titles, text columns, tables, and figures. A grid is exactly what the name implies: horizontal and vertical lines on a sheet of paper the same size as your page that show the locations for all the elements that will appear on the page. A grid determines where blocks of text and illustrations will and can be placed.

Make sure as you develop your layout that the page has plenty of white space—for example, below the headers and above the footers, in the margins and gutters, and between paragraphs. Too much

Pros Know

A nice-to-have feature that enhances any binding method, but is particularly important for large documents, is **tabs.** Tabs are a real boon to usability; they let the user not only see the document's organization at all times, but quickly open to a particular section without flipping through pages or reading headers or footers.

cramped, tight text is unpleasant to read because the eye has to work too hard to make out individual elements. At all cost avoid creating a page that looks like a single, gray block of text.

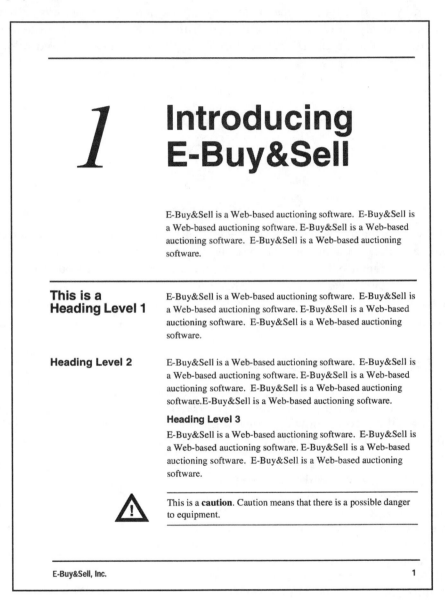

This type of page layout is easy to read and attractive.

Column Width

Choose a column width that makes it easy for your user to read. If your page is large (8$\frac{1}{2}$" × 11" or A4), indent your column width enough so the column doesn't exceed 5$\frac{1}{2}$ inches or so. You can *outdent* your headings at different settings, as shown in the previous figure, or you can align the headings with the text column.

If your page is a small custom size, the column width can be as wide as the margins allow. If you do indent it, make sure that the text column is wide enough to hold tables and other information.

The figure in this chapter of the sample page layout shows the type of layout that places all headings, notes, and icons in a wide left column, with the right column devoted to body text. The top of each heading aligns with the top of the first paragraph in its section. This is a particularly good design if your page is large (8$\frac{1}{2}$" × 11"), because it requires enough space on the left to place the headings.

Be aware that not all word processor and desktop publishing systems can handle this type of page design. Don't choose this layout if it's going to be a struggle to create or maintain it; instead, you can achieve almost the same effect with a single-column layout by strongly indenting the body text so the column is a comfortable reading length.

Color Me Interesting

It's always nice if you can afford to add a second color to your printed documents (other than black, that is), even if it's just a touch in the headers and footers, or to set off major headings.

If your budget or time constraints just don't have room for color printing, you can still add it to documents delivered electronically. Include color in the templates you use to create the documents, then produce the paper documents in black and

Dodging Bullets

When determining your column width, the line of type should be short enough to make reading comfortable. Because the reader's eyes, not head, move when reading, book designers strive to keep column widths fairly narrow: between 5$\frac{1}{2}$ to 6 inches (about 14 to 15$\frac{1}{4}$ cm). The longer the line length, the larger the type size should be, to keep a rough readability proportion between type height and line length.

Tech Talk

Outdent is the opposite of indent. Items of text extend outside the (left) alignment of the text column. This works with headings. It doesn't look very good with numbered or bulleted lists. (We've seen some books in which they've tried this, and then have returned to normal alignment or indenting for later printings.)

white but let color do its thing in your PDFs. Can't decide what color will be your second color? Use one of your company's logo colors as your second color. This creates a subtle but powerful visual connection.

Use color sparingly. Just because you can make every heading, bullet, number, table outline, and figure title in color, doesn't mean you should.

Dodging Bullets

It's tempting to experiment with printing in colors other than boring black, and sometimes you get an effect that's really pleasing. But be wary of depending on colors such as red and green to indicate important differences. Red-green color blindness is not as uncommon as you might think. A good guide for choosing contrasting colors is how they print in black and white—if you can't see the difference between two colors in black and white, choose different colors.

Dodging Bullets

We don't recommend using fully justified text because if you don't hyphenate it, we guarantee you'll see large rivers of white space flowing down the page within the text block. (And you without a life jacket.) If you do hyphenate your justified text, proofread carefully to see that there are no misleading hyphens in the text.

Just My Type: Mastering Typography

The type you use should have clear, clean, open letter forms and should be large enough to read. We recommend *ragged right,* not *justified,* text.

Avoid Hyphenation

For technical documents, including those intended for viewing online, we prefer *flush* left, ragged right text with no hyphenation. Automatic hyphenation creates problems with computer commands, long file names, messages, and product names.

Computer commands should be shown exactly as they must be typed by the user. Most of the time, that means they must be shown all on one line. This can be tricky to capture in printed documents because the computer screen has essentially no limits on line length. Messages also commonly use line breaks to cue readers about something in the message, so they need to be represented faithfully in print and online. Some computer commands and messages actually include necessary hyphens as an element; by hyphenating words, you mislead a reader into thinking the generated hyphen is part of the command or message. There are enough special characters in use in technical documentation without adding another one—the hyphen—to their number.

Type Talk

Typography, like any specialized field that's been around a few hundred years, has its own vocabulary. These terms originated in the days when metal type was used for printing. Because they've been inherited by desktop publishing tools and graphics programs, it's handy to know what they mean.

> ➤ **Font:** The complete assortment of letters, numbers, and special characters (such as punctuation) in one size and one face. This often is confused with *typeface,* which refers to all the sizes, weights (bold or regular), and italic forms associated with any given name such as Arial or Times. So "Arial 10 point bold" names a font, whereas "Arial" is a typeface.

> ➤ *Point* **size:** The size of the font measured from the top of the ascenders (the part that sticks up on a lowercase "h" or "k") to the bottom of the descenders (the bottom of a lowercase "y" or "p"). A typeface that uses longer ascenders and descenders might look smaller than one that doesn't because its letters' bodies might be small.

> ➤ **Leading:** Pronounced *ledding,* this refers to the default spacing between lines of type, and takes its name from the bars of lead that printers formerly used to separate lines of set type. This normally is called the "line space" in word processing programs. Single spacing means there is no extra space between lines:

Tech Talk

Flush: Aligned type (flush left or flush right).

Justified: Type aligned on both sides to form a block-like appearance.

Ragged right: Type aligned on only the left. The right side of the text does not align, giving the block of text a "ragged" appearance on that side.

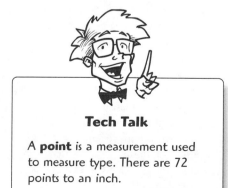

Tech Talk

A **point** is a measurement used to measure type. There are 72 points to an inch.

The bottom of the longest descender in one line reaches the top of the tallest ascender in the line below it. Single-spaced 12-point type is on 12-point leading. To create space (add leading) between the descenders and ascenders of adjacent lines, change the leading to be a number larger than the point size.

➤ **Serif:** The short cross-lines at the ends of the main strokes of the letters in typefaces such as Courier, Times, and Garamond are the serifs. A typeface that has these is referred to as "a serif face." A typeface that lacks these extra lines, such as Arial or Helvetica, is said to be "a sans-serif face." ("Sans" simply means "without.")

Courier, Times New Roman, Garamond, and Arial typefaces.

Courier	Times New Roman
Garamond	Arial

You don't need to reinvent the wheel in choosing a typeface for your documents. There are so many faces available these days you could never see them all, and sometimes differences between them can be subtle. It's hard to go wrong with old standbys. (See the next section, "Cutting to the Chase.") They've been in use for so long because they are what a typeface needs to be: clean, readable, and familiar.

Pros Know

Limit your usage of a fixed–width font like Courier. A fixed–width font (also called a monospaced font) gives every letter exactly the same amount of horizontal space, like a typewriter. This makes the text hard to read and it takes up more space than a proportionally spaced font. Courier is successfully used to represent user input, text file content, computer messages, and similar elements, all of which need a distinctive (and often monospaced) look.

Cutting to the Chase

With all this new information and vocabulary, you might feel as if you can't possibly decide which typeface is the one to use. Relax—you don't have to know them all to choose which one to use. Let's start with the industry standards for different page and text elements:

➤ **Headings, headers, and footers:** Arial (or Helvetica) Bold, of various sizes.

➤ **Body text:** Times New Roman Regular; flush left, ragged right. Add four to six points extra after paragraphs and list items.

➤ **Figures and tables:** Titles in Arial (or Helvetica) Bold, one or two points smaller than the body text size. Text in Arial (or Helvetica) Regular, the same size as the table title.

➤ **Computer messages:** Courier.

➤ **User input:** Courier bold.

➤ **For an 8¹/₂" x 11" page size:** 11- or 12-point Times New Roman Regular with a leading of between 13 and 15.

➤ **7" x 9" or smaller page size:** 10- or 11-point Times New Roman Regular with a leading between 12 to 13 for the 10-point, and a leading between 13 to 14 for the 11-point.

Headings

Decide how many levels of headings you want in your documents. Four or even five are normal, with Level 1 being the chapter title. More than that is too many—your reader will become lost.

Indicate the different heading levels with differences in type size and indents. Level 1 might be 24-point type with no indent, with the successive headings indenting more and more as they decrease in size.

Keep your headings fairly short (two lines maximum, but preferably one) and to the point. A reader should be able to scan the document and get the gist of it from the headings alone.

Dodging Bullets

If you're going to use differences in type and position to indicate different heading levels, make them very different. Make headings distinctive enough that the reader can identify the level even if other headings aren't on the page for comparison.

Make sure your headings are meaningful for the reader who wants to skim the document.

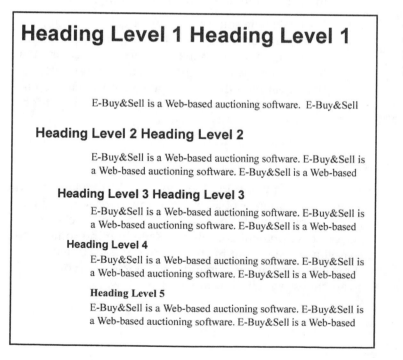

Heading Level 1 Heading Level 1

E-Buy&Sell is a Web-based auctioning software. E-Buy&Sell

Heading Level 2 Heading Level 2

E-Buy&Sell is a Web-based auctioning software. E-Buy&Sell is a Web-based auctioning software. E-Buy&Sell is a Web-based

Heading Level 3 Heading Level 3

E-Buy&Sell is a Web-based auctioning software. E-Buy&Sell is a Web-based auctioning software. E-Buy&Sell is a Web-based

Heading Level 4

E-Buy&Sell is a Web-based auctioning software. E-Buy&Sell is a Web-based auctioning software. E-Buy&Sell is a Web-based

Heading Level 5

E-Buy&Sell is a Web-based auctioning software. E-Buy&Sell is a Web-based auctioning software. E-Buy&Sell is a Web-based

Body Text

You can't go wrong with a standard serif typeface such as Times New Roman or Palatino for body text in printed technical documents. Common wisdom is that in print, serif type is easier to read because the serifs glide the eyes along the tops of the letters.

But computers are changing long-held beliefs in this area, as in many others. So we venture to say that sans-serif type also has its place in technical documents and can be used to good effect.

Users seldom read technical documents from start to finish, all in one sitting. Usable technical documentation also doesn't contain the long blocks of text or long line lengths that serifs helped compensate for; paragraphs generally should be no more than six or eight lines, and often there are visual elements that break up groups of paragraphs. So, many of the old ironclad reasons for not using a sans-serif typeface in printed documents just don't hold up anymore.

Another consideration comes directly from the computer itself. Sans-serif typefaces, cleaner and less cluttered than serif, are now considered to be superior for reading on a computer screen because of the screen's lower resolution. Onscreen, serifs simply add fuzz.

Because single-sourcing documents is becoming increasingly common, some tech writing departments have started using a sans-serif typeface such as Arial for both online and printed documentation. Rather than creating different templates for the same content, they choose the font that works the best for both methods.

Dodging Bullets

Don't use too many different typefaces in a document. They might be fun to play with but the result looks cluttered and is distracting to the reader. When you do use different typefaces, make sure they have a single, assigned meaning, such as Arial for headings or Courier for user input. Avoid using typefaces that are too fancy or difficult to read, like scripts.

Developing Your Layout

When you have some page samples that you think look good, hold a review with the people on your team: other documentation people, engineering managers, and marketing people. Ask for their input, but don't feel you have to incorporate everyone's suggestions—this is not a layout development committee. Keep in mind some of the elements of good design that you've read about in this chapter. If you started by studying documents that look clean and appealing and that people like, you know you have a good base and can choose which ideas are worth following.

Tables and Figures

It's a good idea to use tables and graphics throughout your documents. They support the text, and an illustration can represent clearly and quickly what text alone can't. An illustration can also present abstract concepts in a concrete way. For example, take a look at the figure that accompanies this text:

"A key hardware component in Internet access is the modem. A modem provides the critical interface between the telephone network and a device like your computer or an Internet Service Provider's (ISP) terminal server. Modems make it possible to establish what is functionally an unbroken serial connection between your computer and the Internet, by way of the ISP's terminal server. (See Figure 21-X)"

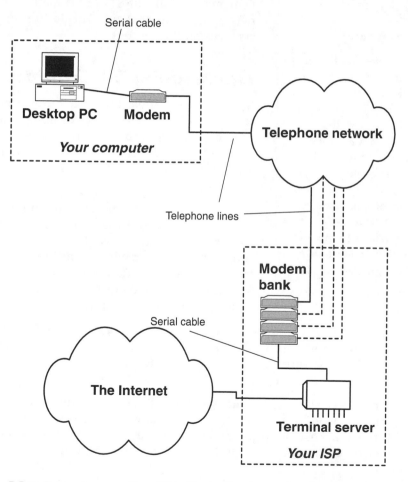

PC-to-Internet connection. This illustration is worth a lot of words!

259

See how the figure adds to the concepts in the text? Maybe *you* were able to visualize everything from the description, but many people require a visual aid in addition to or instead of words. This example illustrates the cardinal rule of illustrations and figures: if they don't enhance reader understanding, leave them out.

Besides increasing understanding by adding a visual element, tables and figures also break up the text. Readers appreciate this because it creates "rest stops" for their eyes as they move down the page. Big, long, heavy blocks of text in a document also are daunting and even intimidating to many readers, and they make it difficult for readers to re-orient themselves after glancing away—you look up once and you can lose your place, and have to start the page over again!

Remember, small "bites" of information and visual variety reduce reader fatigue. This in turn helps the readers focus on the content.

Pros Know

When placing tables and figures in a document, try to limit them to two or three sizes, based on your grid. It's okay if some tables look more full or empty than others—the uniform size of the table outline keeps these differences from being confusing. And make sure the white space above a graphic element is narrower than the space below it—a graphic usually relates to the text above it, and that connection should be visually apparent.

Use Meaningful Graphics

It seems to be standard practice in user's guides to show the user interface screens throughout the procedures. This can be a wonderful help—or totally unnecessary. It all depends on how meaningful each screen shot is, along with how intuitive the interface design is.

If the interface is intuitive and contains complete information, it might not be necessary to show it at all. In this case, you might want to show only the beginning and ending screens, to confirm for the reader that he or she has indeed started in the right place and has gotten the correct result after performing an action or completing a process.

If you do use product windows, make sure there is a reason for showing them. We like the method shown in the illustration that combines text and graphics. The steps of the procedure point directly to the parts of the screen or dialog box that a user is supposed to select or type into.

When using screen shots, make sure you keep all screens at the same scale, so they are in correct proportion to each other. Avoid showing unnecessary parts of the screen. If the difference between one screen and another is important and anything less than obvious, provide some kind of pointer or highlighting to call the reader's attention to it.

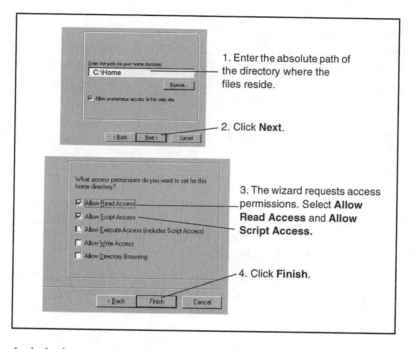

Include the procedure steps with the figure to save space and make it obvious for the user to understand what to do.

Another issue that argues for minimizing the number of screen shots you include is sheer pragmatism. If your product's interface kept changing up until the last minute, there is a great chance that one or more of the published screenshots is wrong! This likelihood increases with the number of screen shots in the document, which is how many you must check and possibly replace each time the interface is modified.

Online Design and Distribution

Many of the principles of design apply equally to online design. The same readers are looking at the documents online, so be sure to still use legible type, plenty of white space, and meaningful graphics.

PDF Documents

If you produce PDFs for electronic distribution, be aware that the settings for online *resolution* and printed resolution are different. A graphic that looks good onscreen can look blurry on paper. We recommend that you choose settings designed for printing; either the user will print the PDF file, or, if reading the document onscreen, presumably has online access to the program and can look at the originals.

WWW Files

When planning Web content, take a look at Web pages you like, those that work for you, and use what you've learned in this chapter to analyze their layouts. Then, take what works and create your own designs based on those ideas.

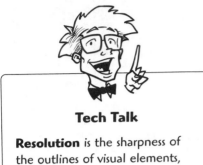

There are many, many books available that can tell you how to design pages for viewing online, and you'll no doubt read differing recommendations about many aspects, particularly what typeface to use. As mentioned earlier, we recommend using a sans-serif typeface for Web design. Serif faces can be very hard to read, particularly on a screen that doesn't have good resolution, while sans-serif faces look uniformly clean and readable.

We've barely skimmed the possibilities of document design; all we were able to do is to provide you with some of the "big-picture" (no pun intended) ideas and vocabulary, so you'll be able to recognize some of these terms when they show up in the Graphics area of your desktop publishing program. If this is an area that interests you, you'll be glad to know that many tech writing jobs consider layout and design part of the job. If this area isn't your cup of tea, well, we hope this chapter can help a little in giving some quick solutions to what could become a big problem—designing a document when you don't have the experience.

Tech Talk

Resolution is the sharpness of the outlines of visual elements, whether screen icons, images, or individual letters in a block of text. High resolution means greater sharpness and detail. Print resolution is much higher than onscreen resolution. Because of this, images or text that appear clear and sharp on the printed page might look fuzzy or muddy onscreen. Keep the delivery medium's level of resolution in mind when planning and creating your documents.

The Least You Need to Know

➤ Ask for help when you really need it—management might wish you were Michelangelo, but you aren't really expected to be.

➤ Learn and use basic design techniques, guidelines, and principles to create an attractive and legible document.

➤ Make graphics meaningful or leave them out.

➤ Remember that good document design means happier users!

Chapter 22

WWWriting for the Web

In This Chapter

➤ The different types of online documents

➤ What makes online documents different from printed documents

➤ About document single-sourcing

➤ Some ways to plan your online document set

A few years ago, it seems all anyone could talk about was how all documentation was going to be online (available via the Internet) soon. "Paper documentation is going to disappear," pronounced the prophets.

Well, it hasn't happened quite that way. Paper documents are still alive and well, thank you. Why? Because users still want documents they can hold, flip through, dog-ear, write on, and read while they're waiting at the dentist's office.

But now paper documents share the stage with online documents, which also have become a key part of what the customer expects—and, therefore, what tech writers have to produce. This chapter tells you how to plan and write technical documents for online or Web presentation. (You'll notice extra "Tech Talk" notes, too—for a new medium you need a new vocabulary, and we supply it.)

Virtual Verbiage

There are many reasons a company chooses to create online documentation in addition to or instead of print documents. At the top of the list is the fact that a paper document that costs quite a bit to print can be generated as a PDF file and distributed worldwide, on demand; through a company's Internet, *intranet,* or *extranet* for little or no cost.

Tech Talk

Intranet refers to a company's internal network, which is based on the same technology as the Internet but is a closed network rather than an open one. Web servers set up on their own internal networks enable employees to use their browsers to access the organization's online documents.

An **extranet** is partway between an intranet and the Internet, and is the portion of a company's or organization's internal network that is accessible to specific outside users (such as vendors or distributors).

The savings in money and time are substantial and obvious. Who needs print shop problems, warehouses, trucks for shipping, or even trips to the post office? A PDF also is easier to update—no need to reprint an entire document or count on users to replace pages. (No worries, either, about what to do with out-of-date manuals sitting in the warehouse.) Online documents get to customers literally in minutes and always arrive in pristine condition.

On top of those strong arguments, online documentation also is more usable and more versatile. If a user has a question while working on an application on a computer, he or she can press a keyboard key (usually F1) or click an onscreen Help button and receive an instantaneous answer. Online help is like instant gratification—users get what they want when they want it, and they want it now. And when customers get the help they need online, that means fewer calls to tech support. Everybody wins.

There's no limit, either, to the number of copies a customer can have for the same cost. Instead of the only copy of a document being locked in the desk of whoever is on vacation, everyone can have his or her own copy—no matter how large or small the office is. With online documents, nobody ever has to be without instructions, reference information, or training materials.

Pros Know

A great addition to your portfolio is a CD of online help that you've created. Take this to interviews so potential employers get the full impact of your work. No CD? Load online help samples to your personal Web site, then give the URL at interviews. One note of caution, however: Be absolutely sure there's nothing at your personal site that you wouldn't want an employer to see!

So what are PDF documents, anyway? We talked about them in Chapter 20, "The Right Tool for the Job," when we told you about choosing tools to create them. PDF documents are something of a hybrid between printed documentation and Web documentation. They're generated from documents created with word processing software or a desktop publishing system, and they look like an image of the document.

PDF documents also contain hyperlinks for easy navigation. These hyperlinks let the user jump from a page number in the table of contents to the page itself, from a cross-reference to the figure or chapter it's referencing, or even from a title reference to that other referenced document.

They're easy to print and give consistently good results. In fact, users often download them for quick access, skim through them, then print either the entire document or only the parts they want. It's true that loose pages aren't as easy to manage as a bound document, but users seem willing to make that tradeoff. (A side benefit to your company is that the user, not your company, bears the cost of printing whatever portion of the document ends up on paper.)

As popular as PDFs are, however, they're only one form of online documentation. In the rest of this chapter, we'll talk about forms of online documentation other than PDFs, because we want to discuss what makes online documentation different from printed documentation.

These are the types of documents you'll read about in this chapter:

➤ Online help

➤ Web-based documents

➤ CD-ROM documents

➤ Computer-based training

Tech Talk

A **server** is a computer in a network that manages network resources. In a network that uses client/server architecture, the server supplies files or services such as file transfers, remote logins, or printing. The **client** is the computer that requests services from the server.

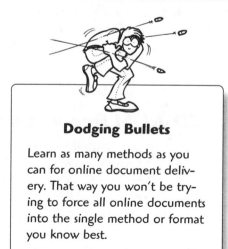

Dodging Bullets

Learn as many methods as you can for online document delivery. That way you won't be trying to force all online documents into the single method or format you know best.

Online Help

Online help refers to any help that is accessible through the computer. It might be part of an *application,* explaining how to use the application as it's running, or it might be a separate document set. Online help is usually linked to a given application. Online help can be browser based (a full set of linked documents viewed through a browser), or it can be *context-sensitive* short help. Context-sensitive help means that the help displayed is related to what's shown on the screen in the area—the context—where the cursor is when you press the Help key or click the onscreen Help button.

Tech Talk

An **application** is a program that enables a user to accomplish a specific task or set of tasks. Application programs are run by system programs (which control the computer) and utilities (smaller programs that perform specific actions, often those of managing a system resource).

A common method of providing context-sensitive help is a pop-up box that displays a few lines of help. Another method, "bubble help," shows a small window like a cartoon dialog bubble that contains short help text. The bubble appears when the user mouses over a button or other interface feature. ("Mousing over" means pointing with the cursor but not clicking.)

Web-Based Documentation

Web-based documentation refers to any document (or set of documents) displayed by Internet technology. It might be on the same computer where the application is loaded or on an Internet, extranet, or intranet page.

You might assume such documents are always produced in HTML (or a similar language), but that's not the case. They also can be produced in many other forms: PDF, multimedia, Word, or other formats. The important thing is that all the parts link together by hyperlinks.

CD-ROM Documentation

Another method for providing online help is on a CD-ROM as part of a product delivery. Like the Web-based documentation, CD help can be produced in any number of formats. Even when this help is browser-based, it can use a browser included with the publishing tool that created the document library, rather than a public browser such as Internet Explorer or Netscape.

Computer-Based Training

Computer-based training has been around since the 1970s but is only now, with the advent of Internet technology, really coming into its own. The form can be anything from a brief tutorial that's only minimally interactive to a full-blown curriculum complete with self-testing, performance reporting, and automatic scoring. What characterizes computer-based training—and is its greatest advantage—is that it can be used at the pace of the learner, and can provide a consistent environment and experience for learners separated by time and geography.

Computer-based training is effective documentation for the customer, but also valuable within your own company. With people changing jobs more often, there's a greater need to train new, incoming employees. It's also often more cost effective to upgrade the skills of a current employee than to hire someone new who already has them. One answer to this need is computer-based training.

When Online Help Is No Help

Even with the best intentions, online help doesn't always help. You know what we mean—you're working along and a dialog box comes up with some cryptic message and two buttons to click: "Yes" and "No." The message makes no sense and you have no clue which button to choose. Perplexed, you press the Help button and up comes the help box. You read it eagerly. What does it say?

"Click Yes to indicate Yes and No to indicate No."

We're sorry if you're smiling and nodding, because that means you've had a similar experience. Too many of us have. This kind of help obviously is no help at all.

Another variation of "help that's no help" is online help that simply pulls up the PDF version of the manual. Repeating the manual word for word doesn't help the user when he or she is looking for a fast answer. When someone is trying to work through an onscreen process or procedure and needs help, he or she shouldn't have to page through the introduction or introductory chapters looking for the relevant part. The user needs focused help for only the process currently underway.

What both of these examples of unhelpful help have in common is that they violate the primary rule of online help: It should provide *added* information that is not obvious on the screen or in another document, say more than the user can tell by just looking, but less than the whole user guide.

Online help also needs to be focused and ideally should fit on one screen so the user doesn't have to scroll. This isn't always possible, but it's a target worth aiming for.

Pros Know

"Tech writers who use human performance technology models are in a unique position to provide input onscreen design, user interface, reports layout, and system help screens. User-advocate design aims at providing consistent operation and visually pleasing presentation. Reminding developers about user vocabulary and requirements reduces the need for training and client support later."

—Connie Swartz, President, Creative Courseware, Inc., www.creativecourseware.com

Dodging Bullets

We've adapted a piece of advice your mother probably gave you and applied it to online help: If you can't say something *helpful*, don't say anything at all.

267

Designing for help needs to be part of product design or Web page design. For example, a main Web page should be very short, with the least information that will convey the idea. This is especially important if it is a page the user visits more than once. In this case, provide more depth and detail about the screen or the page on the associated help page. Help should never be an afterthought—or read like one.

What's Different About Reading Documentation Online?

The single biggest factor that makes reading documentation online harder is that, well, it's online. It's simply harder optical work reading text onscreen than on paper. With paper, light is reflected; with a computer monitor, the "page" is the source of light, and that's harder on the eyes. This means online documentation needs to be shorter overall, and displayed in smaller pieces at one time. People just can't stand to spend as much time looking at an online document as they could reading the same text on paper. (This also partly accounts for why users print out online documentation so frequently.)

If your customers are going to read your documents online, they must be legible. Different from the quality of the writing, legibility means the text itself can be easily seen and the characters interpreted—on a screen with any resolution. The pages also must be short enough so that users have to scroll little or not at all. Colors must not be jarring—no purple text on a pink background, please, even if those *are* your company's logo colors—and color must not be the sole difference between two similar choices or items.

Aside from being legible, online documentation also must be optimized for the medium. These days that means constructed in a modular way so that parts read out of sequence still make sense and still relate to other parts. Online documentation isn't like a book or a movie: Users don't start at the beginning and continue on until they see "THE END" or reach the back cover. Users move around within help screens in whatever way seems good to them, following impulse as well as logic, and seldom in a linear path that's more than two or three screens long.

Coffee Break

In August 2000, just over half of American households had computers, up from 42.1 percent in December 1998. 41.5 percent of all homes had Internet access, up from 26.2 percent in 1998.

Online documentation is like a big network of connected highways and the users need to know which exits to get off to find the ways to their destinations. If a user is just cruising around on a vacation, the places stumbled upon should be interesting. And, like Hansel and Gretel, users should be able to find their way back when they need to.

What's Different About Writing Online Documentation?

Like thinking in a foreign language, your mind has to work in a different way when you write for delivery online. That's because of the nonlinear nature of the way people use online documents. Here are "The Big Seven" guidelines to help you "think online":

1. **Group online information into "chunks,"** each of which should stand on its own. A "chunk" can be a page of text, several pages, a whole document (such as a PDF), another Web site, or a graphic. The size of a "chunk" depends on the overall size of the complete body of information you're working with. And you can't assume in one part that a user has read any other part. If the user needs to have read something else, say so. Then provide a link to the other place and tell the user to click it.

2. **Make information chunks easy to get to.** And, naturally, they also must make sense to the users when they get there. Provide a thread of continuity by carrying key phrases from one page to others that link to it. A reader should never wonder "Why am I here? What does this page have to do with my question?"

Dodging Bullets

When creating online documentation, treat every chunk of information as if it is the only one the user will ever see. In many cases, that really is true!

3. **Make information short and clear.** People reading online documents are usually doing so because they are in the midst of a task and need immediate help with it. Stick to the point and be brief. Put descriptive or explanatory information in a clearly marked separate file. You might even link to it from here, but don't include it. The question "What do I do next?" is much more common, and nearly always more pressing, than "Why do I have to do this?" or "How does this work?"

4. **Keep illustrations to a minimum, and make every one count.** Tables, matrixes, graphs, diagrams, charts—these can be useful (as long as they support what the user is trying to do online). Screen shots, however, although common in print manuals, are often out of place online. The user is at the screen and doesn't need to see an exact duplicate of what's up there now.

5. **Consider the needs of all your users.** The content of the online documentation should appeal to every level of user. Write basic information for the user who needs less handholding, and supply links for users who need or want more, such as detailed instructions or background information.

6. **Clearly differentiate varying types of information.** The user needs to be able to tell, at a glance, what is background information, what is conceptual, what is procedural, and so forth.

7. **Give your users immediate feedback.** Assume that they are working with an application at the moment they are reading your text. Tell them what the result should be and where they go from there. If there are common results that indicate a common error with a known correction, be sure to provide those, too.

Single-Sourcing

Before you start writing your online documentation, you should understand what *single source* means. No, it's not a place to shop for unmarried people. It's a method by which you can take one document file (the single source) and use it as the basis for all your document needs: print documents (including marketing materials), online documents, and online help. Single-sourcing takes a lot of planning, but the results definitely can be worth it. As with other high-tech processes, time invested on the front end can pay big dividends at the back end.

The "foreign language" metaphor applies to single-sourcing as well as to online help. Instead of writing a linear document, starting with page 1 and continuing to the end, you must think of your entire information base as the chunks of information (like we discussed earlier in this chapter). Each chunk is a building block; it might be a paragraph or a group of paragraphs, a chapter, or a whole document. It might even be a single sentence. But each one exists on its own.

Pros Know

Single-sourcing, although it takes a lot of work up front, can save up to 60 percent on documentation costs over the life of the project. If you can reduce documentation costs to less than half of their current required outlay, management is bound to sit up and take notice.

The goal is to construct all the final documents you need from this collection of building blocks. For example, instead of massaging a "procedure" building block one way for a user's guide, another way for online help, and a third way for a training guide, you use the same "block" in every end product that needs a procedure. The only difference is how to choose and arrange those building blocks to create new structures.

Here's another example: The user's guide might contain five description blocks, 12 procedure blocks, and some other miscellaneous blocks. The online help might contain those same 12 procedure blocks but no other shared files. A marketing brochure might contain three of the same description blocks and two sales-pitch blocks. Yet all were created from a single source.

Single-sourcing eliminates the problem of keeping information up to date when it appears in more than one document. Because there is only one source file for each idea chunk, that single chunk can quickly and easily be revised as the product evolves.

Single-sourcing also is valuable for translations. A translator simply translates the source file one time, and you arrange the same information into the final result.

If you're charged with producing multiple types of documents, we recommend you look into single-sourcing and do your best to implement it, even in part. You can start out slowly, breaking the source into chunks of information and arranging the chunks into Web-based documents. Then, gradually make the chunks smaller and smaller.

Getting to WWWork

In Chapter 20, we discussed the tools you need for creating online help and Web-based documentation. You should have some way to create HTML documents, and a help tool that lets you build a document library with a linked table of contents.

The process for creating Web-based documents is similar to that for creating print-based documents except in one significant area: you allocate your time differently. Remember in Chapter 6, "Five Steps to Creating a Technical Document," where we talk about what percentage of time is spent on the documentation? Here it is again:

➤ 50 percent for gathering information

➤ 10 percent for planning

➤ 20 percent for writing

➤ 10 percent for verifying

➤ 10 percent for redoing

When you create Web-based documents, plan on a different time breakdown and order. Both implementation and testing take much more time than they do for creating print documentation. If you already know about the product, the percentages are more like the following:

➤ 40 percent for planning

➤ 30 percent for implementing

➤ 20 percent for verifying

➤ 10 percent for redoing

The percentage of your time devoted to testing and redoing is much greater with Web-based documents simply because of the medium. Implementation is a bigger portion of the overall job as well.

Planning

To talk about planning an online document project, let's use a hypothetical case. Imagine you are producing online documentation for a game. During your planning, you decide you will need to cover the following topics:

➤ Description of the game.

➤ Rules of the game.

➤ Instructions for setting up the game and the players. (This page links to details about the different levels of play and links to three different pages about how to play the game with one, two, or three or more players.)

➤ Options for customizing the game (colors, wallpaper, sounds, and so forth).

➤ Troubleshooting.

➤ Copyright and company information.

➤ How to also play the game on the Internet.

Like any type of documentation, consider the audience for your game documents and what the people in your audience need to know and do. Make sure you have information for all levels of expertise (is this a player's first time?) and identify the different levels clearly.

Tech Talk

Drilling down refers to writing or using online documents that contain levels of increasing detail and complexity. At the "surface" or highest level is the least complex information. If a reader wants to know more, he or she "drills down" by following a link to a page with more details. Want even more? Follow another link. In this way readers who need different levels of information can all be accommodated.

As you think about the topics to cover for the game, plan your information topics. Show the big picture first and make links to the smaller parts. This also is known as *drilling down*.

When developing your online documents, think about how information is structured online—how the pages will relate to each other. Think three-dimensionally instead of in a straight line! Your online documentation is like a tree, with each limb growing branches that in turn produce their own branches. You must control the branching to create a logical structure. The paragraphs that follow describe a typical document structure you might use for your game.

There is a top-level page (let's call it the home page) that shows a graphic depiction of the game. This home page has six links:

➤ Description of the game

➤ Rules

➤ Setup

➤ Options

➤ Troubleshooting

➤ Copyright

Each of these links goes to a separate page. Each of these six pages also has a link *back* to the home page even if there are no other links (never let a page be a *dead end*).

The Copyright page also has a link to your company's corporate Web site. (This reminds you that online documents don't have to be limited to only the pages you actually create.)

Let's take a closer look at the Setup page, which has five links on it (aside from the link back to the home page):

➤ Levels of play

➤ Rules for a single player

➤ Rules for two players

➤ Rules for three or more players

➤ Playing on the Internet

Each of these five pages contains links back to the Setup page and back to the home page. Don't make users retrace their steps if they don't want to (there's always the Back button for those who do).

The page about playing on the Internet also contains links to an Internet Web site. This demonstrates the anticipation of other things the user might want to do or know and takes advantage of the unique capability of an online document.

Can you visualize the page structure we've just described? Probably not—keeping track of all these parts and pages, and which one links to what other ones, is hard to remember. And this is a *simple* documentation set! Even though tools such as FrontPage show you a page in a structural context, it helps to draw it up for yourself (especially during the planning phase). A white board is great for this because it's easy to make radical changes; just make sure when it's finished you copy it before somebody erases it!

Tech Talk

Dead end is the name for an online page that has no link on it leading to another page. The only way the reader can leave a dead end page is by using the browser's Back button. Don't create dead end pages; link back to the home page if you can't think of where else to go.

Dodging Bullets

If your company is one of the many moving toward a "paperless office" by implementing online documentation and converting existing documents to this new medium, remember that sometimes paper documentation is still appropriate and necessary. For example, installation instructions should always be available on paper, as should certain troubleshooting advice. The user won't be happy to learn that instructions for "What to do if your computer crashes" are only available online.

If you don't have access to a white board, don't be shy—get some flip charts or even rolls of brown wrapping paper to tape up on the walls in a conference room. (Psssst: Children's washable markers are great for color-coding links and topics.)

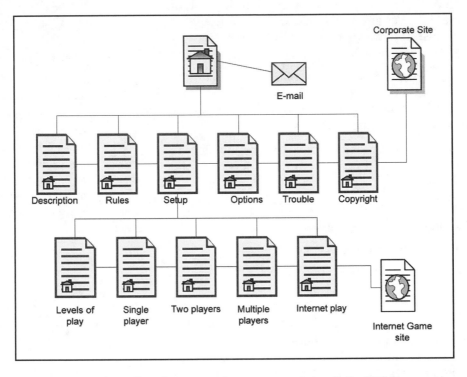

This flowchart for online documentation corresponds to all the HTML pages in the document set.

Break your information into chunks, and assign a category ("procedure," "description," "task-based," and so forth) to each chunk. Then go at it, placing the chunks into pages and connecting them in sensible (and creative) ways.

Implementing

In developing online documents, implementation always takes a larger percentage of your allotted time than for print documents. So, be sure to keep this in mind and plan for it.

One technique that will help you keep a handle on your files is to put all of them into a single folder and give each file a name that helps you identify its place in the overall structure. For example, try names that indicate the category, the type of chunk within the category, and an assigned order or level number. A name such as *proc-install-1.html* would mean it is the first chunk that is an installation procedure.

274

Don't, however, rely on memory or an "obvious" naming scheme: Create an explicit "key" that identifies the parts of the name and explains what the different values in each part mean (for example, that "proc" stands for "procedure").

If you start feeling overwhelmed by the sheer number of pages you're working with, you might feel tempted to start cramming more information on each page. Don't make this mistake. Keep information in short, easy-to-digest bites, especially at the higher levels. This forces you to constantly ask yourself how much is enough and to stay within that focus.

Verifying and Redoing

As you work, plan to test frequently. Open the files in your browser and check the look of the page and the links. Check that the information flows in a logical way and that illustrations are correct. Make sure no pages are dead ends.

You'll find that your online pages rarely are right on the first try. It's always something—three visual elements are out of alignment, a link isn't working, a heading is bigger (or smaller) than you intended. But don't let this discourage you; it's part of the development process, and the reason that testing takes so much time.

The only way to make sure your documents work like they should in the end is to create a test plan. We don't mean something haphazard and casual; we mean one that can be used by somebody outside your group. The test plan should ensure that every button, every link, and every menu choice works as intended. It also should test the documents on Internet Explorer, Netscape, and any other required browser. Things have a disconcerting way of looking and behaving differently on different products.

Also be sure to test for usability as well as plain function. Can the user understand the information easily? Can the user find the information he or she is looking for? Can the user navigate through the infrastructure without a problem? The answers to these and other usability questions must be "yes" before you can consider your online document to be finished and ready to deliver.

Online documentation and single-sourcing are entirely new concepts to many of us, and not easily tackled on the first try. But well-structured documentation can be well worth the extra effort it takes to plan it up front. Luckily, gurus like JoAnn Hackos, Janice C. Redish, and William Horton have written some of the bibles for these subjects, and you'll find their books indispensable as you start on this path. See Appendix B, "For Your Bookshelf," for the titles that can help you.

Pros Know

Online documentation requires much more testing than printed documentation does. The reason for this is obvious: printed documents do, after all, just lie there. Your online documents must perform.

The Least You Need to Know

➤ Embrace online documentation—it is more than the wave of the future; it is the way of today.

➤ Design online help to make the user's life easier.

➤ Implement single-sourcing for maximum effectiveness and long-term cost savings.

➤ Plan (and draw) your document's online structure and page interaction in detail before starting to work.

➤ Don't try to re-invent the wheel; this is a tough subject. Read some of the books by the experts and follow their advice.

Part 5

I Love My Job, I Love My Job, I Love My Job ...

In this part, we look at tech writing from a few angles we haven't presented yet.

We wouldn't be telling you the whole truth if we didn't say there will be days when you'll have to chant "I love my job, I love my job, I love my job ..." to yourself as a reminder of why you keep coming back to work every day. Tech writing has its own unique set of challenges and frustrations, no matter what company you work for.

Some of these frustrations can be reduced by taking advantage of new ways of getting work done outside a traditional office setting. These introduce their own complications, however—solvable, but still needing attention.

But even if you sometimes have to remind yourself how much you love your job, it probably will be true more often than not that you do love your job. So, we wrap up the book with insights and advice about planning where you want your tech writing career to go.

Chapter 23

Office Alternatives: Working Outside the Box

Bob Dylan wasn't thinking about the work environment when he sang "the times, they are a-changin'," but if he had been, he'd have been right. Not only do people no longer expect to stay with a single employer until retirement, now they don't even assume they'll have to come to the office every day. More and more people are exploring different employment models; the constant in many workers' lives has become the consulting company that farms them out instead of the office where they spend their workdays (which could change monthly).

The whole relationship between employers and workers has begun to change, and we think it's going to keep on doing so. In the information economy, people are the most important resource. Add to that the one-two punch employers take every day in employee commute time and the cost of office overhead, and it's not surprising that everybody is exploring new ways of getting work done. In this chapter we take a look at work options you might be considering—or maybe never thought of before. Kick off your shoes and get comfortable.

Consultant or Captive?

Perhaps the first question is whether it's really necessary to keep working for somebody else at all, or at least whether the company that pays you has to be the same one where you keep your coffee mug and favorite photos. And this isn't a clear one-or-the-other question, either, as you'll see in this chapter.

Different people use the terms consultant, contractor, and free agent (or freelancer) in different ways. We offer some definitions to these terms, although in different parts of the country and in different companies the distinctions blur. The real similarity between these workers is that they are not permanent full-time employees of the companies for whom they work; in fact, they might and often do perform work for several firms in a given year.

You Say Consultant, I Say Contractor

In a very general sense, a consultant is an independent worker who comes into a company to solve a problem or work on a project. The consultant might be a full-time employee of a consulting company or might be an independent. Often consultants receive a set fee for a project, since they are selling a solution, not their time.

A contractor fills a job temporarily, on a contractual basis. The contract sometimes defines the length of time and payment per hour rather than charging by the project. That way, the contractor is able to function like the permanent employees do, working on-site and on the same projects for the same number of hours. Many hiring managers and high-tech workers use "contractor" and "consultant" interchangeably, although general consensus seems to be that the consultant is in more of an advisory position, hired as a problem solver, while the contractor provides services. Some people joke that a consultant is a contractor who gets paid more.

The Best of Both Worlds

Both contractors and consultants might be employees of a consulting company (sometimes called "job shop" or "outsourcing" company). These types of companies keep some full-time staffers and call in contractors when they need to staff a project for a client. Such relationships can be convoluted: One writer we know was employed by a West-Coast outsourcing company that farmed her out to another West-Coast consulting company, which then put her to work on a project at a third company located in her home town in the Midwest.

If you work as a consultant (or contractor) for a company, you often have benefits such as paid vacation and sick time, health insurance, even savings plans and paid training. In some ways, this is the best of both worlds: You get the job diversity of a contractor with the benefits of employment. The company places you on projects at their client companies, sometimes short-term or sometimes on jobs that last for years. If there are no projects available at client sites, you might work at the consulting company's home office. Your hourly rate brings you more take-home pay than if you were a permanent employee at any of the companies you work for.

But there's also the downside of being an employee: Your job is to fit into the client's corporate culture, whatever that happens to be. If you enjoy the variety of working for many different clients but you like the steadiness of continuous employment, you'd probably do fine in this capacity.

Some consultants and contractors work on a different basis. Rather than receiving a regular paycheck, they must invoice someone to get paid. If they are subcontracting with a general contractor—perhaps one that finds the projects, manages bookkeeping and client relations, and bills the client—their invoices go there. It's common for general contractors to form close alliances with subcontractors and clients and to be in those relationships "for the long haul."

If a contractor is working directly for the client, or if he or she is the general contractor, the invoices go to the accounts payable department at that company. Independent contractors and consultants can be paid by the hour or might work based on a set fee for a project.

If contractors actually are self-employed, they are on their own in securing health insurance, planning for savings, and setting aside the right amount of money for taxes, not to mention making sure that the next job is lined up when the current one ends. But they usually have a correspondingly greater say in their everyday activities—such as whom they work for and what hours they work. Even when a subcontractor works almost exclusively with one general contractor, there's still a sense of autonomy in the relationship that's important to many people. If that's important to you and the prospect of finding your own work doesn't daunt you, contracting on a self-employed basis might be something for you to explore.

Coffee Break

Who telecommutes? It might not be who you think:

"According to studies published in 1997 by the U.S. Department of labor, more than 70 percent of persons who did some work at home in 1997 were in married-couple families. It isn't just women with kids to take care of who are working at home. The D.O.L. study shows that women and men were about equally likely to work at home and that the work-at-home rate for married parents was about the same as the rate for married persons without children."

—Janet Attard, "Who Works at Home?" (www.businessknowhow.com)

Free Agents

Free agents (also called freelancers) are the "mountain men" and "mountain women" in the world of contractors, going their own highly independent ways. Self-employed or incorporated, they might juggle three to five different clients in a single week, on top of marketing their services and handling bookkeeping. Even if they have a few core clients, they're more likely to work on short-term one-shot projects than to seek closer continuity.

This is the ultimate in tech writing entrepreneurship, and definitely not for everyone. But, as one free-agent writer we know, who works with high-tech *and* low-tech companies, said in a recent e-mail, "Hey, it's not just everyone who deals with turkey slaughter houses, diamonds, drag racing, and electrical test systems for the F-16 all in one day!"

The "Bottom Line" Is on a Tax Form

What tells the ultimate tale is taxes. If you work on a "W-4 basis," you're an employee, even if you're *called* a consultant or contractor. The W-4 is the name of the government form you fill out when a company hires you. It identifies the company as your employer and specifies the amount of taxes the employer withholds from your paycheck and pays to the IRS for you (along with some of their own money). This arrangement is the one people nearly always mean when they talk about "having a job."

If you work on a "1099 basis," you're some variety of free agent (even if you have stable, long-term relationships with contractors and clients). The 1099 is a government form that a company sends to anybody *not* an employee who they've paid more than a certain amount of money during the tax year. They also, of course, send it to the IRS. In this arrangement, the client withholds nothing—the dollar amount on the invoice is what they put on your check. Taking out a chunk for taxes is up to you.

And speaking of taxes, we urge you to consult someone knowledgeable about the tax implications of these new ways of working if you're considering any of them. There's no substitute for up-to-date information from someone whose job it is to know the federal and local tax laws that apply to you.

Dodging Bullets

Don't toy with your taxes. If you've recently started contracting or working as a free agent, check with a tax consultant about how much to withhold and how often to pay your federal and local taxes.

Can You Really Work Without Getting Out of Your Bathrobe?

The short answer is "yes." We don't know how many hours we've worked over the last few years in a favorite terry bathrobe and fuzzy slippers, but it's been quite a few.

Before you go shopping for loungewear, however, there are some things about working from home that you need to understand.

Discipline? Me?

First, it takes discipline. Many people find the idea of working from home appealing precisely because it removes the office constraints. But it also removes the office structure and the office motivation. Maybe nobody will see you if you spend 20 minutes looking through your daughter's kaleidoscope, but that doesn't mean that's the best use of your time. Because there's nobody around to "crack the whip" over you, you have to do it for yourself—or the end of the day comes and you have little to show for it.

A Time to Rest and a Time to Work

You also have to be resistant to distractions. That doesn't mean you can't throw in a load of laundry or watch television when you take a break. It does mean you have to resist the temptation to spend time on home projects that beckon to you just because

Pros Know

Succeeding as a professional in the high-pressure world of high-tech industries doesn't just mean working extra hours every week—it also means knowing when it's time to take care of yourself. You can't keep putting in those 60-hour weeks indefinitely. Trust us on this one. There will always be more work than you can get finished. Take time out before exhaustion, or the flu will do it for you.

you're right there. This also means resisting children, pets, and neighbors who might think that because you're home, you're available.

The flip side of this is knowing when to stop. With your work always right at your elbow or in the other room, it's easy to slip into being "always on," especially when your job is an especially demanding one. Maybe your body is at home more now, but your family will notice that your attention isn't. Is your life really only about your job? When you're working from home, it can start to feel that way.

That "Hermit" Feeling

There's no way around it—working at home is lonely. Maybe you're not close friends with everyone at the office, but those interactions are still human contacts. Everybody needs those; human beings are social creatures. It's true that some people are more comfortable working alone than others but even the most introverted or self-sufficient people still miss saying hello to people in the morning and goodbye to them at night.

When the Walls Start Closing In

"Cabin fever" is another ill you might encounter. It's that feeling that if you don't get out of the house you're going to scream. The wallpaper pattern grates on your nerves, the broken miniblinds in the kitchen taunt you, and if the neighbor's dog barks one more time you don't know *what* you might do. You can't remember the last time you really "got dressed," and the loose threads and pet hair on that bathrobe are really starting to bother you.

All These Problems Have Solutions

There are rational, effective solutions to all these problems. For example, schedule lunches with people from the office a couple of times a week. That gets you dressed and out of the house, and keeps you in touch with your colleagues. Having a bad attack of cabin fever? Take 10 minutes and walk around the block, or make a quick trip to the grocery store to pick up some things for dinner. Often all you need is a brief change of scenery to get yourself back on track. As for knowing when to stop, the clock can be your friend at home even if it felt like your enemy at work—start and stop at the same time each day, and when you're not in your office, close the door. Create and maintain a clear boundary between work and home.

There's Still No Place Like Home

With the right balances in place, working from home can be great. Walking down the hall is the shortest commute in history. There's nothing like that smug feeling of looking out your home-office window at ghastly weather and knowing you don't have to go out in it. It's okay to take 30 minutes when the kids come home from school to greet them. And we have to admit there's an undeniable sense of fun in conducting serious, professional business meetings in a bathrobe or sweatclothes.

Our point is that working from home isn't all room service and a day at the beach. But it can be wonderfully liberating, if you go into it with your eyes open. There are many books and Web sites available about the growing trend toward working from home, and we recommend reading a few of them. They'll help you decide if working from home is for you.

Pros Know

There's lots of information and advice about telecommuting on the Internet. Here are a few sites to get you started:

➤ telecommuting.about. com/small business/ telecommuting/mbody.html

➤ www.telecommuting.org/

➤ www.telecommute.org/

➤ www.tdigest.com/

About Telecommuting

There are really two varieties of *teleworkers:* someone who is a regular employee who works from home, usually only a few days a week but possibly full-time; and someone who is a contractor whose home office is the nerve center of his or her self-employed life.

Regular employees who telecommute usually don't have to supply their own equipment. Typically the employer provides the computer and peripherals, just as at the office, along with basic office supplies (such as printer paper). Bear in mind that all this equipment still belongs to the employer, even though it's in your home. (Some family members might need an occasional reminder.)

It's not unusual for the employer to also pay for high-speed data connections such as xDSL lines or

Tech Talk

Teleworker is a generic term beginning to be used along with "telecommuter" for anyone who works at a site other than the company's main office, including from home or from another telework site. Teleworkers can be either regular employees or contractors.

at least a second phone line. Getting employees access to a company's internal networks can be a challenge, but more solutions for this are becoming available all the time; and that one's up to the employer—not the employee—to solve.

The contractor setting up a home office is in a completely different boat. The cost of everything comes out of your pocket (although you can usually write it off your taxes as a business expense; talk with your tax consultant about that). Because of this you might be tempted to go with a bare-bones setup that will save you hundreds of dollars. Think twice. You need equipment that's on a par with what your clients are using, because you have to be able to load and use the software they've chosen.

You also must be able to work with the same rate of productivity as people in-house; you don't want anything—such as a low-end microprocessor or not enough RAM—to slow you down. So bite the bullet and buy a system that's good enough for today and that can be upgraded readily tomorrow.

What are the basics? A computer, of course, with a decent-sized monitor. You don't need something that will give you a tan, but it does need to be large enough to fit most of a page on it—and you still must be able to read it. A printer is a necessity, of course. When choosing between a laser and an inkjet think about the kinds of things you'll be printing and how often you'll need to print in color.

You also might want to add a flatbed scanner; you'll find it very handy for working with graphics. You may or may not need a free-standing fax machine; it's easy to send and receive faxes straight from your computer, but some fax machines double as an *Optical Character Reader* (*OCR*) scanner and feed-through photocopier and can be very handy if you need those secondary features.

Tech Talk

OCR (Optical Character Reader) is a technology that reads printed text documents from paper and converts them into text files (not images of the pages).

Ultimately the clients you work with, as well as the kinds of work you do for them, drive your equipment choices.

How to Work Off-Site and Still Be a Team Member

If your company is one of the "early adopters" of new business practices, there might be as many people working off-site as on-site. Team members might even be scattered around the country. If that describes your organization, this topic won't be as much an issue for you.

But if you're one of the trailblazers at your company in working from home, you might have to deal with "out of sight, out of mind." It's important that you address these issues proactively to keep yourself in the loop:

➤ **Communicate frequently and regularly.** We mean real-time voice conversations, "the next best thing to being there." This means making sure you can get in touch and stay in touch with colleagues quickly and easily. A hands-free phone is a must, and making it also cordless is even better. Get at least three-way calling and call waiting on your voice line, along with caller ID. Learn conferencing software such as NetMeeting—you might even need to introduce it to your boss. Making sure others can reach you when they need to, and can expect a prompt or instant reply, helps them forget you're off the premises.

➤ **Become expert with your e-mail.** Make sure your e-mail reader checks the mail server every few minutes, and alerts you when new messages have arrived. If you're away from your desk or office a lot, it might serve you to invest in a pager that can beep or buzz when you get new mail. Acknowledge messages promptly, even if you can't immediately answer them fully. Manage your e-mails, too, by sorting and storing them so you can find a particular one later.

➤ **Provide status reports before you're asked.** Tell your boss (and colleagues, if it's appropriate) how things are going without being asked. If you used to seek their input and feedback on things, keep doing so now. Let them see that nothing has changed except where you're sitting.

➤ **Go to the office.** Working off-site is not the same as being let out of jail. Be willing and cheerful about coming in for meetings, and drop off documents in person once in a while. Take a look at what's posted on the lunchroom bulletin board or by the coffee machine. If you're athletically inclined, you might even want to join the corporate soccer or softball team.

A transition time between being on-site full time and being off-site full time helps everyone get used to the new arrangement. Start by working from home one or two days a week, but don't make those days both Monday and Friday. Too much crucial business gets done in offices on those days for you to miss both of them, at least during the transition. Being "gone" both those days also can create the perception that you're working a short week, even though *you* know you're busy at home. Choose one or the other, to start off, or go with midweek days only.

Maybe you'd like to try telecommuting, and feel confident it would work in your situation, but you don't know how to bring up the idea and sell it to the people who need to agree. The key is to show what's in it for the company, not just how much you'd like it. There are plenty of places on the Internet that can give you advice about making a case for telecommuting—they anticipate the questions management will ask and include convincing statistics along with solutions to logistical problems. Besides the URLs listed in this book, a search on the term "telecommuting" will return abundant hits to get you started. Showing you've done your homework and are aware of the issues from management's side of the table increases your chances of success.

It takes genuine effort, commitment, and buy-in from management, but office alternatives such as teleworking really do, well, *work*. Maybe you need more than one bathrobe after all.

The Least You Need to Know

➤ Understand what's involved in being a consultant, contractor, or free agent so you can best choose which might be right for you.

➤ Think creatively to address the headaches of working from home so you can fully enjoy the benefits.

➤ Know when to say when—have clear hours for working and clear hours for not!

➤ Be proactive about keeping yourself in the loop once you're working off-site.

➤ Get your ducks in a row before approaching management about the idea of telecommuting—show what's in it for *them,* and the company, too.

Chapter 24

You Didn't Think It Would Be Like *This!*

In This Chapter

➤ The downsides of the tech writing field

➤ Some tips on taking responsibility for yourself so you aren't overwhelmed during deadline time

➤ Information about proprietary materials: what you own and what you don't

---

Every job has its difficult aspects, and tech writing is no exception. Much as we love tech writing as a career, we know only too well that there are hurdles in it that you won't find in any other field.

In corporate-speak, they aren't problems, they're "challenges"; and we have found that how you look at them, think about them, and respond to them makes a big difference in whether they make you crazy. In this chapter we balance an honest presentation of some of the challenges with workable solutions that can minimize the impacts on your day, your year, and your career.

The "Dark" Side of Technical Writing

Professionals in today's high-tech industries face a number of problems along with the pleasures. It's not all flight simulators and free pizza with extra cheese. The excitement and fast pace of the computer industry also mean unrelenting pressure and the stress that comes with it. The tight deadlines and heavy demands of release dates inevitably mean long hours of relentless work staring at tiny characters on a computer monitor.

As if your tired, scratchy eyes weren't enough, today's desktop publishing programs more and more depend on menus and dialog boxes, and can cause repetitive stress injury from too much mouse use. *Dilbert* cartoonist Scott Adams wasn't kidding when he quipped in a cartoon panel that technology was no place for wimps.

Choose Your Deadline: Aggressive or Insane?

Those don't sound like very appealing choices, but they're often the ones you get to pick from. It's a hazard of the computer industry, not of tech writers specifically: By nature all deadlines in the high-tech industry are … aggressive. No industry in history has ever moved at the speed of this one. A tech writer of 15 years ago might have counted on having six months or a year to write a single manual and release notes, and probably didn't have to worry about addressing document formatting or graphics.

A tech writer in an Internet company today, for example, will tell you in a matter-of-fact voice that he or she might have six *weeks* to write a similar manual and release notes. This writer won't even bother to mention that he or she also has to do the graphics and produce camera-ready copy as well. And that's probably not the only thing on that writer's plate—it's probably in addition to writing or editing a few other documents—and, oh yes—some online help!

This kind of intense pressure, coupled with the lack of control that many tech writers feel, can lead quickly to *burnout,* and we're talking "puff of smoke on the horizon." Job burnout can result in a loss of interest in the quality of your work, inability to meet the demands of the schedule, and feelings of intense exhaustion on all levels. The only effective way to avoid burnout is to acknowledge that it's a factor and deal with job stresses before they occur. We've tried in this chapter to give you some ideas of how you can take matters into your own hands when it's possible, and to stress less when it isn't.

Documentation Always Comes Last

This is an observation as well as a complaint. In the first place, nobody can write about something until it exists, at least on paper, so in the natural progression of things, the bulk of the documentation can't come anywhere but last. (We've heard of companies that let tech writers write the documentation first and make the programmers build the product to match, but we suspect these actually are only urban legends started by tech writers.)

In the second place (this the part you won't like), when product managers and marketers plan a product, they don't always remember to tell the tech writing group. More times than we can count, a developer has nonchalantly asked if we knew about the upcoming product release. (Sometimes only a few weeks away!)

Even when the writers are aware of the documentation needs and willing to work on-schedule, the developers still might not be available to provide help and information. Often, the developers work steadily (sometimes frantically) on a project until it is complete. Only after they submit their deliverables do they realize that they have neglected the tech writer, who by this time probably has left them 100 e-mail messages and started waiting to ambush them in the parking lot.

There are some ways you can deal with this fact. Try to do as much as you can ahead of time. A tech writer who gets involved in the product development process near the beginning always saves time at the end. It sounds obvious, but many writers do not get an early start even when they get early notification. Most, if not all, of the documentation can be written and completed before the product is complete. Why? Because the developers know what the product is going to do. In fact, they often have specifications that tell them exactly what they're expected to produce. It's a process of inference to write the documentation from the specifications.

Dodging Bullets

No specifications? Then ask the product developers what the product will do by the time of the release date. Don't base your documentation plans on what the product is doing while it's still in the development stage, or what the final, ideal "dream version" of the product is. Document what reliably will be. Remember, you wouldn't like all *your* colleagues to assume your draft is going to be identical to the final product!

When You've Got to Pull a Rabbit Out of a Hat

Many developers and tech writers don't dare document something until it has actually been completed by a developer and sometimes even tested. Sometimes this caution is warranted, but usually not—this often causes you to lose valuable time.

What about when you're asked to write a manual for a product that doesn't exist yet, has no specs and no prototype, and the one developer assigned to the project is too busy to speak to you?

When you're stuck with what must surely seem like an impossible task, find out what the product is going to do and, if possible, how it works. If the product has a similar function to one that already exists in your company, create an outline based on an existing document and do your best to fill in the parts that you think will be repeated in the new document. Once you have these placeholders, you can fill in the information as you receive it.

Afraid that the team's best-laid plans might not materialize? Write it anyway. It's a lot less stressful to remove or slightly modify some material at the eleventh hour than it is to spend that time writing full documentation on features that somebody knew were going to be in there from the outset; yet this happens all the time.

Changes, Changes, Changes

The one thing that tech writers complain most about probably is the fact that so many changes come in at the last minute. Because the tech writer does so much work at the end of the project, he or she often has to deal with a shower of requests for changes at the last minute. It can be very frustrating: developers come to you and ask you to make a change; then after you make the change they come back and have you change everything to what it was before! It can start to feel like an ongoing conspiracy.

It is frustrating, and if you let yourself feel powerless because of it, it can cause a huge amount of stress. Remind yourself that everybody is a victim of the constant changes—not just you or your department—and instead of taking it personally, just relax and accept it as part of your job. If you think of "doing major changes at the last minute" as part of the job description, something you expect and can handle in a competent and professional way, you'll feel much less like a victim.

The developers and the rest of the staff are suffering from the last-minute changes, too; it's not just the tech writer. They are finding errors and fixing them as well, often on a scale that would make your palms sweat; this whole "deadline" idea isn't much fun for anyone.

Take Control

These solutions to the last-minute chaos might not be perfect, because you can't always control what other people do. But you can control what you do, and there are a lot of ways to overcome some of the madness that tech writers have to go through. Most of these solutions involve you taking matters into your own hands and becoming active from the start. They are more than just workarounds.

Here is a list of solutions—tape them up on your office wall if that's what it takes to remind you to use them.

1. **Get involved early** in the product life cycle. Get the support of your manager if he or she is not aware of the problems and state clearly that you want to be involved in projects from the beginning. It is very unlikely people will say "no"— they usually like it when a writer participates in project meetings. Attend all the meetings related to the project you're working on. You'll all hear what's happening with your product at the same time, and together you can anticipate problems with the documentation. People also get used to thinking of you as part of the project team.

2. **Don't be a hermit.** Part of the reason tech writers don't hear about what's going on is because sometimes they isolate themselves from the people in the thick of things. Make a point of being part of the team. You will hear what's being planned, what's being released, and what's in the wind. Whether this means attending meetings, going to lunch with people outside your group, or talking while making coffee in the break room, make a consistent effort to do it.

3. **Make sure that your company's product cycle process gives a nod to documentation.** If your company has a formal release cycle process, ask your program manager to see that one step in the process requires the product manager to plan early for product documentation. Having it in writing as an official part of the process will at least make your colleagues aware of the need to plan and request documentation at an early stage, and—surprise!—many of them will include the writer in release meetings, too.

4. **Take steps to become independent.** Tech writers are too often dependent upon developers for information and assistance. Anything you can do to reduce this dependence is to your advantage, especially during the last-minute schedule crunch. Independence can include becoming involved with QA testing and having access to a system where you can check pilot software against your document.

Coffee Break

What's your personality type? The Myers-Briggs Type Indicator and Keirsey Temperament Sorter determine what type you are based on four categories: Extrovert and Introvert, Intuition and Sensing, Thinking and Feeling, and Judgment and Perception. "Rationals" make up only 5 to 7 percent of the population, but it's been suggested that most technical people such as engineers and tech writers fall into this group. Take the Keirsey Temperament Sorter test at http://www.keirsey.com.

5. **Take the initiative.** Don't say, "I'm waiting for the programmer to give me information on E-Widget's new features." Instead, write the text yourself, as well as you can. Guess, if you have to … or lie. Then take it to the programmer and ask him or her to check your information. If your facts are wrong, the programmer will be quick to correct you. Enjoy being incorrect—it will help you get your work done more quickly and keep the programmer from being irritated that he or she is doing "your" job.

6. **Do what you can to relieve stress.** There are many problems you won't be able to solve no matter how much initiative and energy you have. For those problems, we say don't waste any more time on them than you have to. If you know you can't do anything about it, there comes a time when you must give up, accept the facts, and do what you can to take care of yourself. Exercise regularly, don't eat a lot of junk food, and try not to overload on caffeine. You probably already know the drill—but you have to *do* these things, not just think about them or complain that you don't have the time.

"I Don't Get No Respect"

Feel that the tech writer is the lowly peon in the hierarchy? You aren't alone. It's a common complaint in many tech writing departments. The writers feel that they are looked down on by the other members of the technical team.

Is this true? Sometimes yes, sometimes no. If you feel there's a lack of respect for your technical ability, you have a couple of choices for how to deal with it. First, assess whether this is the truth. If it is, change it. If it's not true, it might be a problem with an individual, or it might be the corporate climate.

It's true that in some companies, tech writers are seen as little more than glorified word processors or administrative assistants. If that's true in your company, the solution might be to get what you can out of the job, do your best, build your portfolio, and use that experience to move upward and onward to something better. If this company doesn't appreciate you, look for another that does. You'll find one.

Tina the brittle technical writer in Scott Adam's *Dilbert* is the personification of the tech writer who is quick to take offense at the way she's being treated. Of course, sometimes she's right to take offense, as the cartoon shows.

If your company does not expect its technical writers to take all the minutes at meetings and you see that the other tech writers don't feel a lack of respect, the challenge is to improve yourself professionally. Think about what it would take for you to feel on a par with your peers. Take classes that enhance your technical skills and product knowledge. These classes are going to be a help to you wherever you go, so don't feel you're "selling out" to the people who look down on you. Make sure you aren't making a lot of errors or doing careless work. Make sure you're not expecting developers to do your work. You should be doing the writing; they should help by providing information about what they are working on and by giving you technical reviews.

The world's most famous tech writer, Tina, is sometimes justified in her fears. (But we love her for putting tech writing in the public eye!)

(DILBERT reprinted by permission of United Feature Syndicate, Inc.)

In the long run, it's whether you are a first-rate technical writer that's important, not whether you are a second-rate programmer. Any technical skills you acquire are going to help you to become a better writer. And that's best for you, and for the company.

Working with Problem People

Yes, we know. Everyone, everywhere sometimes has to work with problem people. Why is this a special problem for the tech writer?

As a tech writer, you must depend a lot upon other people to help you do your job. As we said previously, many different developers, QA people, marketing people, customer service reps, and salespeople are responsible for providing information to you. Unlike some workers in the computer industry, you are responsible for dealing with a lot of different people. And you have to maintain a pleasant relationship with those people, because they give you what you need.

It's a little different for the ones who are providing the information. At some level, they are calling the shots. Yes, they have deadlines, and the product delivery often is dependent on what they say and do. They have the goods and everyone else, tech writer included, wants those goods.

And some of these people are hard to work with. Sometimes it's just that they don't seem to know you exist. They don't return your e-mail. Or they won't read your draft copies. Or they're never around when they've promised to help you learn something about the product.

Other times they can be genuinely not-nice people. And if, under the stress of deadline pressure, you overreact to one of these types, it can blow up into a problem that can hurt your career.

Ask your manager for help with the situation before it escalates. Make sure you cover yourself by leaving a paper trail: Send e-mail to people you don't trust (actually, it's a good idea to keep a record of everything; you often need to refer to it later or use it to prove something) and spell out in detail everything you plan to do with or will need from that person. Send a copy of the e-mail to your manager, or if possible the problem person's manager.

If someone at work is preventing you from doing your job and you have a detailed record that shows that to be true, ultimately all you can do is the best you can. Make sure your difficult colleague's name is on the sign-off sheet in big bold type; include a line next to the name for him or her to sign, saying that the documentation is okay. Not willing to sign? Fine. If there's a problem with the documentation, one name will be conspicuously missing from the sheet. If you can't complete a document because of someone, turn the problem over to your manager.

Finally, remember that it's not your fault if someone else is a difficult person. There are ways to deal with each type of problem person—and ways that help ease the stress on you—but there's not a lot you can do to change their personalities.

The Complete Idiot's Guide to Getting Along with Difficult People, by Brandon Toropov can help you find some methods of handling yourself under these difficult situations.

You're Doing All the Work While Someone Else Gets the Credit

If you're a person who loves to write, you might be a little disappointed by the realities of tech writing life. Do you mind writing a book that won't have your name on it? Can you handle losing "ownership" of a book that you labored on so lovingly? Will you be unhappy when the next writer changes your elegant prose or rearranges the sections you thought about so carefully?

Those are difficult hurdles for some people. At some point, as a good tech writer doing your best to create a quality product, you have to learn to think of the document as just that: a product. It's not the Great American Novel.

Who Owns the Documentation?

As a paid employee or contractor of a company, you are producing *work for hire*. This means that even if you do all the work, the company that pays you is the owner of your output and therefore can do what it wants with that work. This means that your company can give your document to a subsidiary company, a customer, or anyone they choose, and let that recipient fold, spindle, or otherwise mutilate it.

How Proprietary Is Proprietary Information?

Tech Talk

Work for hire refers to work paid for—and therefore owned—by one person (the employer) and created by another person. The employer is the legal "author" of the work, and therefore owns the copyright.

When you start work at a company or contract with a company to produce work for hire for them, you sign a nondisclosure agreement (NDA). The NDA states in essence that the company owns the work you do and you may not show it to anyone else without permission. It also probably says that you cannot do any work for a competitor.

Read the NDA before you sign it. (Some tech writers don't even remember signing one; to them, it was one of many forgettable administrative papers signed on the first day of the job. To them and to you we say, "read the manual!")

Consider how the contract can affect you: If you write proprietary documentation for the company, you are never permitted to show it to anyone else. Even user's guides are considered by many companies to be restricted material because they are distributed only to a limited set of paying customers.

The day will come when you want to change jobs. And to do that, you are likely to need writing samples. We've heard of tech writers who take samples clearly marked "proprietary" to an interview. Some interviewers don't even object … at least you think they don't; of course, that might be the reason they don't call you back for a second interview. Some interviewers not only object, they have been known to take the document from the interviewee and call the company to return their proprietary information to them. That doesn't sound like a promising interview!

Getting Samples for Your Portfolio

When you look for a new job, the interviewer probably is going to want to see writing samples. This is not the best time to go to your employer and ask permission to show your samples!

The best way to handle this situation is to face it when you accept the job. Tell the hiring manager that a condition of your accepting the job is that you will get samples of the pieces for your portfolio. Work out the restrictions under which you will be allowed to show them. Some employers will agree that you can have them as long as you don't let them out of your sight or permit photocopying. Others ask you to use only certain parts of the documentation.

You don't want to go empty-handed to an interview where the hiring manager expects to see samples. What will you tell that manager—that every piece of work you've ever done was proprietary? You will likely need to explain all the details of the jobs you've done. Most interviewers like to have a sample to look at as you explain what you've done. If they can read it and associate your explanations with the sample, the impact stays with them longer—especially if the document is from a well-known company or product, or is a standout visually.

If you cannot get permission from your employer, try this method: Disguise part of the documentation you wrote by putting another company and product name in there, changing anything that would give it away. Just be careful about how you present it: If you've only worked at one company and your disguised documentation is about a product that happens to do just what your company did, that's cheating. The first time, it's not exactly a secret whose product it is. After you have a few jobs under your belt, that will be a fine portfolio piece.

The Light Side of Tech Writing

All in all, technical writing is not so dark. It's a satisfying job that offers you many opportunities to learn, create, and deal with some pretty cool technology. And the current demand means that you have a good chance of finding the job you want and leaving it if you're not happy. Dark side? Nothing a bright tech writer can't deal with!

The Least You Need to Know

➤ To ward off deadline pressure, write the document as early as you can, even if the developers haven't put all the features in.

➤ Learn enough about the product so you're not dependent upon the developers to do your work.

➤ Remember to treat yourself right—exercise and eat healthy foods to help keep the stress at bay.

➤ When you take a new job, make sure you get permission to use some of your work in your portfolio.

Chapter 25

Managing Your Career

In This Chapter

➤ The different levels of technical writers

➤ Tips for how to keep your knowledge up to date

➤ Advice on networking (the people kind)

The daily life of a tech writer has a lot of variety, but it's variety within certain boundaries. There comes a time in most writers' lives when they want something to change. It might be more money or more challenge and responsibility; it may even be a move in a completely different direction.

This book has been about getting in the tech writing door, and getting up to speed once you're there. We've now come to the end of the book, but you're just getting started. It's time to talk about what to do when you've built some solid experience as a professional technical writer. If you've followed the recommendations in this book, you're well on your way. This chapter is our send-off, to ensure your success continues. Good luck!

Where Do You Go from Here?

Cast your imagination a few years into the future. You've been working as a tech writer for a few years now. You're doing good work and you enjoy your job, but your honed tech writer's mind keeps asking one question: What next?

The answer depends entirely on you. Many tech writers enjoy what they do so much that they don't feel a need to change it. For many it's the perfect balance of structure and autonomy. You work independently most of the time and do a variety of work every week, even every day. You interact with a lot of interesting and intelligent people and participate in exciting development work.

But everybody's different, so some writers become restless, feeling as if there's no place to go from where they are. It's true that tech writers in the computer industry don't have a lot of upward mobility. That's because of the unique nature of the work we do. For instance, if you're part of the engineering department, it can be hard to get a promotion because you're not an engineer.

One way to move in your career is to become a consultant, contractor, or free agent, as we talked about in Chapter 23, "Office Alternatives: Working Outside the Box." But if you're not the type who wants to fly solo, there are other paths to explore: Look for ways to move up or move laterally within your current company. And there's always the possibility of going to a different company altogether. We talk about those choices later in this chapter.

Let's start by looking around where you are. There probably is a career path for tech writers in your company. Aside from the salary grade levels defined by your company's human resources department, there probably are standard job levels in your department, too. Here are some ranks of tech writers that we've seen at various companies, with a brief outline of what might be expected from someone at each level.

A Junior Technical Writer Job Description

A junior writer is an entry-level tech writer. At this level, employers expect you to be able to update and edit documents, proofread, and do some writing under supervision. Nobody should expect a junior writer to write an entire new manual from scratch without plenty of guidance. Junior writers need—and should expect to receive—learning time to develop expertise in the product, the tools, and the company's standards. After all, that's what "junior" is all about.

A Technical Writer Job Description

Writers at this level typically are what a hiring manager has in mind when looking for a tech writer. A mid-level technical writer can work with minimal supervision and could reasonably be expected to create a serious document from scratch with only a little guidance. This writer knows how to follow a schedule and plan work to meet deadlines. The mid-level technical writer also is proficient in the tools needed to do the job, and has enough experience to pick up new tools quickly or even choose or recommend tools for a specific task.

sounds like SBL 2003

A Senior Technical Writer Job Description

This is the top of the line for tech writers in a nonmanagement position. You now are a seasoned professional—"The One Who Knows." A senior tech writer also is called a lead writer because he or she might be expected to guide or mentor other writers with less experience. The senior writer can manage him- or herself fully independently and can confidently take a project from concept to finish with no guidance.

Pros Know

What makes you a senior writer? One consultant we asked had this to say:

"A Senior Writer is a writer who can think, not just react.

➤ Can you take a project from concept and turn it into something you can hold in your hand?

➤ If you are given a deadline, do you know how to plan the 237 little steps involved so that when the deadline arrives, you've hit them all?

➤ Can you sit among your company peers and when the lead developer says, 'We need to have an API document created because we're adding a Developer's Toolbox to the application,' say 'Okay, I can do that' (and you don't even know an API from a hole in the ground)?"

—John Posada, Senior Technical Writing Consultant

In addition to sheer experience, a senior writer has some extra abilities above and beyond the mere job requirements: mastery of more than one publishing tool, for example; as well as extra knowledge such as programming skills, graphics skills, and tool knowledge; and some other expertise such as telecom, hardware, or networking background. By this point in a career a senior writer is likely to have preferences and specialties, but can write any type of documentation on demand. Needless to say, these are the writers nearly every company is hunting for.

Documentation Manager

Some people might think that managing is managing and that any manager could run a documentation department. Not so. A documentation manager needs to know the document development process from hands-on experience. Theory isn't enough. Usually that's not a problem because most documentation managers came up through the ranks as writers. In many companies, the documentation manager also is expected to be a working writer; that's seldom a problem because most managers still like to keep their hands in the game anyway.

But you will need additional skills if you're aiming for a management position. The duties of a manager fall into four categories: planning, organizing resources, leading, and coordinating (note the absence of "writing" even though you'll still do that, too).

Aside from whatever writing he or she takes on, the manager is primarily expected to plan the workload for the department and develop schedules that coincide with product deliveries. He or she has to meet deadlines that often are driven by some other group, and must manage the time and effort of everyone on the documentation team to meet those deadlines.

A manager also is the liaison between the documentation people and other departments, as well as upper management—no small task in itself. And, on the other side, a manager is responsible for making sure his or her group has the resources it needs to get things done. This involves anticipating department needs and getting requests through the proper channels so those needs are met on time. (That new printer won't help you if it arrives two weeks past the deadline you needed it for.)

Managing the department also might mean managing the department budget—you'll need to be good at juggling numbers, as well as those flaming sticks we talked about.

Moving Up

In case it isn't already clear, let's say this right out: A career in tech writing is not about opportunities for climbing a clearly defined corporate ladder. The advantage, the benefit, even the appeal of a tech writing career is in the opportunities it offers for continuous learning, varied daily activities, and a high level of autonomy (as well as making a good living).

Depending on your areas of interest and personal goals, you may not be sure what to do once you reach senior writer level. Not everyone wants to be a documentation manager, and there's great wisdom in knowing whether this path is for you. And not every company offers a separate career path, that of "technical expert," which some seasoned tech writers can step into.

Nevertheless, don't assume there's nowhere for you to go within your current company. Some large companies have separate documentation departments that do not work for any one area, but are called upon to work on projects for any department or division that needs what they can do. If you're not in that department already, find out if it's a change you'd like to make. That also might be the avenue to advance into management, if that is where you'd like to go.

Think also about the people you most enjoy working with. How might you enter their ranks? Writers have more transferable skills than almost any other kind of worker—skills such as analyzing, organizing, and communicating. What department in your company doesn't need people with those abilities? Everything else can be learned.

Moving Out

In tech writing, just as in other fields, sometimes you simply "can't get there from here." When that's the case, the only solution is to go somewhere else.

Don't feel as if you've failed if this seems like your best choice after only a relatively short time in this field. The average time at a job for tech writers in some of the high-tech areas is less than a year and a half. You won't be the only one who started looking farther afield after only a few years on the job.

Luckily, the demand for technical writers is great in many parts of North America—and indeed, nearly everywhere in the world. We won't give you standard job-hunting advice here; that's available in lots of other places. We will, however, talk about working outside your own country. English is becoming the de facto universal business language and tech writers who also are native English speakers have a definite advantage in that arena. Major corporations are continually expanding their worldwide presence; sometimes product documentation is produced only in English even for products marketed worldwide. Israel and all the European Union countries routinely hire English-speaking tech writers to create their English-language documentation.

A motivated job seeker definitely can find work in another country and have a rewarding experience. You'll use all your skills, tech writing and otherwise, mastering a new language and a new culture in your daily life even though you might speak lots of English at work. You'll deal with engineers who don't speak any English at all, and you still have to get the information and determine whether it's right. Sometimes the only skill a new hire is checked for is to be a native English speaker, which means your on-the-job experience could be almost anything. But if you're up for the adventure and have ever wanted to live and work in Europe, the Middle East, or nearly anywhere on the Pacific Rim, tech writing can be the ticket to

Coffee Break

Be aware of cultural differences in a foreign country job hunt. The rules truly are different there. As an American, for example, you'll be surprised to learn that European employers can legally make age a hiring criterion, and might use psychological tests, IQ tests, and even handwriting analysis to profile applicants for hiring decisions.

your dream. An exciting life? Absolutely. We've included some Web sites in Appendix C, "Professional Organizations and Web Sites," to help you get started in your search for overseas employment.

After reading those last two paragraphs you might feel like a shoo-in for an overseas job. Maybe you are and maybe you aren't. Getting a job in another country is not automatic, even if you have all the skills an employer needs today. All countries have rules about employing nonresidents, and the employer usually must meet certain requirements, such as proving they couldn't find a qualified resident. If you work longer than three months in another country, you will need a visa in addition to your passport, which can take quite a long time to get. If you want it, the patience and persistence needed to land an overseas job can be worth it. But, as with any job change, don't quit your current job until you have a solid green light from your next one.

Moving Into Another Field

Because the people who make good tech writers also are people who need to keep learning new things, a lot of technical writers change careers after a few years. It isn't because tech writing necessarily becomes boring (though for some it does). It's just that motivated tech writers in the computer industry get the chance to explore such a wealth of other topics; that's one of the beauties of the high-tech business. You can start in one place and happily wind up somewhere else.

We know people who have used what they learned and the connections they made as tech writers to become system administrators, usability specialists, project managers, product managers, marketing managers, trainers, and programmers. We even know one tech writer who went into human resources. Pretty much everything that you write about within the high-tech business can be a doorway into your second—or third, or fourth—career.

Keeping Your Knowledge Current

The future for a tech writer indeed is wide open—but we're getting ahead of ourselves. You still have to make it through those intervening years, and jobs, before you can feel like the world is your oyster. To make yourself more marketable in the short run and to give yourself more choices in the long run, always keep your knowledge current. A tech writer with a busy schedule often forgets to think about what skills he or she should learn next, or what knowledge might be getting obsolete.

For the most versatility, enhance your technical skills. (Hmm, doesn't that sound familiar?) Take classes in the field you work in (telecommunications, networking, semiconductors) or in programming. Choose languages that already have solid bases of programs written in them, rather than the newest Java-wannabe that may or may not take off. Updating existing programs might be your transition into a new field, just as new tech writers start with updating existing documents.

Take classes related to the area that interests you, but also keep an eye out to acquiring skills that will transfer readily into other areas. If you want to learn more about the business end of things, study marketing or business. If you want to learn more about project management, take some classes in that, or at least subscribe to a *listserv* or follow a newsgroup about it. That's good advice anyway, regardless of what path you want to pursue.

Don't neglect to keep your skills sharp in tech writing–related areas, too. Try to keep up on what's happening with topics such as single-sourcing, on-line help, Web-based documentation, HTML, SGML, XML, and graphic design. Not only can you get inspiration from seminars and workshops, but you'll also meet people who are useful contacts.

Once you develop some expertise in the field, consider teaching or mentoring newcomers into the field (yes, you!). Community colleges and tech writing certificate schools (often classified as vocational schools) offer many opportunities to teach what you live every day. It can be a satisfying experience, offer variety in your work, and—not incidentally—give you first pick of the up-and-coming new technical writers in your area as they step into the working world.

Pros Know

Consider entering some of your organization's technical documents into the Society for Technical Communication's annual competitions. You can submit your entry to a chapter contest; winners of those go on to compete in the international competitions. Choose from categories such as Online Communication, Technical Art, Technical Publications, and Technical Video. Get more information at http://www.stc-va.org/fcomp.htm.

Networking Doesn't Stop Just Because You're Employed

Network, network, network! We said it earlier when you were job-hunting, and we're still saying it now. Continue to make contacts in the field during your entire working life. And networking doesn't mean just with other writers. People you meet on a project who work in other fields can turn into good contacts later.

Even if you never leave your present position (and how likely is that?), networking is beneficial. Knowing other people in the tech writing field helps you when your company needs to hire someone, for example. (And if your company gives a bonus for bringing in a new hire, so much the better!)

Networking also helps when you need to find a better way (or any way) of doing something or solving a problem. If you can call on a pool of experienced writers, you can save yourself crucial time and effort, especially in a crisis.

If the time comes when you do need to seek outside opportunities, your networking contacts are the ones you'll turn to first.

Join Professional Organizations

You're unlikely to bump into a bunch of technical writers at your local grocery store and talking to your co-workers, although enjoyable, doesn't exactly count as "networking." The best way to meet colleagues is the same way you met people who were doing what you wanted to do "way back when": join professional organizations (maintain your membership if you already took our advice!). Appendix C has a list of several. You're almost sure to find something you like there.

Go to STC Meetings

The Society for Technical Communication (STC) is the biggest organization (25,000 members) for technical writers, and they have lots of chapters in the United States, Canada, and even around the world. Go to http://www.stc-va.org/fmbr_menu.htm for information on how to join. Then start going to your local chapter's meetings. Work on chapter committees or hold chapter offices, if those are things you enjoy and are good at. They look great on your resumé, in addition to being good ways to build your list of contacts and keep on top of new developments in the profession.

Tech Talk

A **mailing list** is an e-mail discussion group, also known as a **listserv** or just a list. All members receive the list simultaneously, and when you respond on the list, every member sees it. You can choose to receive the postings as either individual e-mail messages or in digest form (24 hours' worth of postings consolidated into a single message). For more about mailing lists and how to subscribe, go to http://www.lsoft.com/lists/listref.html.

Join the TECHWR-L Mailing List

TECHWR-L is a monitored but unmoderated discussion forum for technical communication topics. This Internet *mailing list* has nearly 5,000 subscribers worldwide and the discussion is pretty lively. The list is designed for those working, or interested, in technical communication jobs.

TECHWR-L also has its own Web site, the Official TECHWR-L. The Web site, at http://www.raycomm.com/techwhirl/, is full of all kinds of technical writing–related information, with regular columnists writing on an array of topics. The site contains archives, too, of all the TECHWR-L mailing list postings. Got a question about tech writing and wondering what your peers would say? Check this Web site and the archives for all kinds of information you'll need and use, whether you subscribe to the list or not.

Although your life as a tech writer might be so rewarding that you never need to make a major career change, it's nice to know that you could. Start preparing now for your future by exploring all the paths that are open to you. The things you do to help yourself today are what will make your working life better tomorrow—and for years to come.

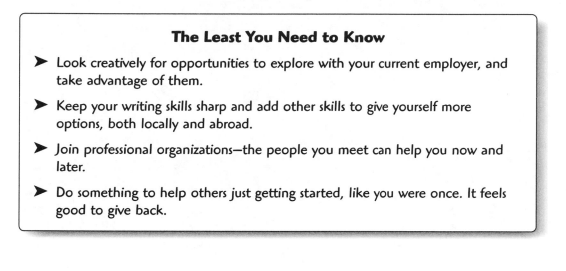

The Least You Need to Know

➤ Look creatively for opportunities to explore with your current employer, and take advantage of them.

➤ Keep your writing skills sharp and add other skills to give yourself more options, both locally and abroad.

➤ Join professional organizations—the people you meet can help you now and later.

➤ Do something to help others just getting started, like you were once. It feels good to give back.

Tech Talk—
The Tech Writer's
Glossary

API (Application Program Interface)   An interface between the operating system and application programs.

application   A program that enables a user to accomplish a specific task or set of tasks.

bio-break   Although not a true technical term, it's one you'll be glad you know. Use it when you need to ask tactfully for a break in a meeting or interview to "answer the call of nature" or attend to other biological functions like the need for a drink or snack ("How about a bio-break, everyone?").

bitmap   Any of several data structures that represent information—typically images or a set of characters—as a collection of bits.

bridging words or **phrases**   Also called "transitional" or "linking," these are words and phrases that indicate a connection or other relationship between two or more paragraphs. Examples of bridging words and phrases are "additionally," "on one hand/on the other hand," "consequently," and "meanwhile."

burnout   A state of physical, emotional, and mental exhaustion, sometimes marked by apathy or strong resentment, usually caused by long-term work situations that are stressful and demanding.

camera-ready   In print production, a camera traditionally shoots each page or set of pages as the first step in making the plates that run on a printing press. Camera-ready art or copy is, literally, picture-perfect and ready for production.

change requests   Official requests, usually submitted through a formal internal system or process, for error (or "bug") correction in software or documentation.

contractor   A professional who works on a contract basis, often as a subcontractor for a general project contractor and usually for a specific project and with a budgeted number of hours.

copyright   The exclusive ownership of a literary work, a musical work, or a work of art that gives the right to make use of the work, protected by law for a specific period of time.

dead end   This is the name for an online page that has no link on it leading to another page. The only way the reader can leave a dead-end page is by using the browser's Back button.

Desktop Publishing Tool (DPT)   A program that lets you use a computer to create camera-ready copy for printing. DPT combines the capabilities of word processing programs, page layout programs, and drawing or painting programs. Adobe FrameMaker, Adobe PageMaker, QuarkXPress, and Corel Ventura are some common DPTs in use today.

draft   Any iteration of a document before the document is finalized.

drilling down   This term refers to writing or using online documents that contain levels of increasing detail and complexity. At the "surface" or highest level is the least complex information. If a reader wants to know more, he or she "drills down" by following a link to a page with more details.

emoticon   A combination of characters, usually punctuation or other nonletter characters that, when viewed sideways, suggest facial expressions.

end user   The intended user of your product or document; differentiated from the people who designed or created it.

entry point   In programming, a place in a program where execution can begin. Expanded to apply to documentation, an entry point is any point or avenue a reader can use to access a body of information. An index is a collection of entry points, as is a table of contents. Subdivisions of a document also are entry points.

extranet   Partway between an intranet and the Internet, an extranet is a portion of a company or organization's internal network that is accessible to specific outside users (such as vendors or distributors).

FAQ (Frequently Asked Questions)   Pronounced *fak* or *eff-ay-kyew;* abbreviation for a list of questions commonly asked by people new to a subject and answers to those questions. FAQs originated on newsgroups and often are part of Internet sites and mailing lists as well.

flush   Aligned type (flush left or flush right).

Graphical User Interface (GUI)   Pronounced either *gooey* or *gee-you-eye;* this acronym refers to an interface that uses graphic icons and conventions such as dialog boxes, icons, multiple windows, and pull-down menus in addition to words.

gutter   The white space on the edge of the page toward the binding. Also refers to the space between columns.

hard copy   Document printed on physical paper, whether from a laser printer or an offset printing company.

hyperlink   A link in an HTML document that jumps the user to somewhere. Hyperlinks usually are underlined or shown in a different color from the surrounding text. They can be pictures as well as words. See also *hypertext*.

hypertext   Online text that contains a hyperlink.

index marker   The symbol generated by the word processing or desktop publishing program that indicates where an index entry is.

information transfer   A term from the science of library and information management that describes how information "moves" from its source to become integrated into the personal knowledge stores that an individual uses as a basis for actions, decisions, and judgments.

intranet   A company's internal network based on the same technology as the Internet, but that is a closed network rather than an open one.

Java   An object-oriented programming language developed by Sun Microsystems, Inc. Similar to C++ but easier to use, Java is a popular language for writing small applications, called "applets," used in pages on the Web.

JavaScript   A WWW scripting language developed by Netscape Communications and Sun Microsystems, Inc., which is loosely related to Java.

justified   Type aligned on both sides to form a block-like appearance.

mailing list   An e-mail discussion group, also known as a listserv or just a list.

margin   The white space on the outside of the image area (top, bottom, and outside on a bound page, or all four sides on an unbound page).

OCR   Abbreviation for *Optical Character Reader*. Technology that reads printed text documents from paper and converts them into text files (not images of the pages).

Open-Source Software (OSS)   Software in which the program source code is openly shared with and among developers and users.

outdent   The opposite of indent. Items of text extend outside the (left) alignment of the text column.

peer editing   Editing done by a colleague who is at an equal standing within a company or organization, rather than farther up the hierarchy.

person   In writing, a way to indicate how near the reader (or writer) is to what is being said. The first person is "I," "me," or "we," the second person is "you," the third person is "he," "she," "they," or "them."

production   Carried over from the days when everything was printed on paper, this term now stretches to include many complex processes. For example, it might mean taking your letter-perfect text files to printed, bound, shrink-wrapped manuals delivered by the hundreds at your company's warehouse. Or, it might mean morphing those files into interactive pages stored in a database integrated with scripts, graphics, and other software to "publish" them on your company's Web site.

proper noun   The name of a *particular* person, place, or thing. Proper nouns must be capitalized.

QA (Quality Assurance)   A set of procedures that includes the entire development process, monitoring and improving the process, making sure that agreed-upon standards are followed, and ensuring that problems are found and resolved.

ragged right   Type aligned only on the left. The right side of the text does not align, giving the block of text a "ragged" appearance on that side.

reference   (Also called "locator.") The part of the index entry that takes the user to a place that answers his or her question. It is a page number or set of page numbers, or a pointer to another subject, typically indicated with "*See* topic XYZ."

resolution   The sharpness of the outlines of visual elements, whether screen icons, images, or individual letters in a block of text. High resolution means greater sharpness and detail.

saddle-stitched   A printing production term for a binding method in which individual pages, folded in the center, are laid over a stapling machine (as if they were a saddle on it) and stapled (stitched) together.

server   A computer in a network that manages network resources. In a network using client/server architecture, the server supplies files or services such as file transfers, remote logins, or printing. The *client* is the computer that requests services from the server.

single source   A noun meaning a file (or folder full of files) from which *all* product documents come, created by selecting and rearranging different pieces of the source into different information configurations—the output documents.

single-source publishing   The ability to create all your user documentation, specifically printed and online, from a single original file.

SME   Pronounced *smee,* this is a common tech writing acronym for Subject Matter Expert. A SME can be anyone from a product manager to someone working in shipping—it all depends on the subject matter expertise you need.

soft copy or **electronic copy**   Document that is read from the computer screen.

stakeholders   The people who have a stake in seeing that a given document is accurate and complete.

stock   The term printers use when talking about any kind of paper, from the lightweight paper you use in your laser printer to heavy, coated paper used for covers.

teleworker   A generic term beginning to be used along with "telecommuting" for anyone who works at a site other than the company's main office, including home or another telework site.

time to market   The amount of time it takes to get a product on the shelves so it can start making money.

topic   (Also called "subject.") The word, phrase, or abbreviation listed in alphabetical order in the index. Together with the reference, topics make up the index entry.

trigger   The action or state of being that initiates a process.

usability   The practice of taking the physical and psychological requirements of human beings into account when designing programs and documents.

voice   A grammatical term that describes how the subject and verb in a sentence relate to each other. Active voice means the subject is *doing* the action of the verb. Passive voice means the subject is *receiving* the action of the verb (with no indication of who or what is doing the action).

white paper   An industry term for a paper (such as a report) that usually states a position or proposes and explains a draft specification or standard.

white space   Space on a page that doesn't have type, graphics, or anything else in it. White space shows up on the page as margins, the gutter, the spaces between headings and paragraphs, and the open space around graphics or at the end of a chapter.

work for hire   A work paid for (thus owned) by one person (the employer) and created by another person. The employer is the legal "author" of the work, and therefore owns the copyright.

workaround   A temporary solution to dealing with a bug or other unresolved problem that enables users to "work around" it until it's fixed.

YMMV (Your Mileage May Vary)   One of many abbreviations used in informal written computer correspondence.

For Your Bookshelf

Although we've drawn almost exclusively on our own experiences in writing this book, we know that you'll want to refer to other books for information during your tech writing career. Here is a list, although not exhaustive, of books we've found useful.

Alred, Gerald J., Charles T. Brusau, and Walter E. Oliu, *Handbook of Technical Writing.* New York, NY: St. Martin's Press, 2000.

Bondura, Larry. *The Art of Indexing.* New York, NY: John Wiley & Sons, 1994.

Cooper, Alan. *About Face: The Essentials of User Interface Design.* Foster City, CA: IDG Books Worldwide, Inc., 1995.

Corcodilos, Nick A. *Ask the Headhunter: Reinventing the Interview to Win the Job.* New York, NY: The Penguin Group, Plume, 1997.

Edwards, Paul and Sarah. *Working from Home: Everything You Need to Know About Living and Working Under the Same Roof.* Los Angeles, CA: Jeremy P. Tarcher, Inc., 1990.

Galitz, Wilbert O. *The Essential Guide to User Interface Design: An Introduction to GUI Design Principles and Techniques.* New York, NY: Wiley Computer Publishing, 1997.

Gordon, Karen Elizabeth. *The Transitive Vampire: A Handbook of Grammar for the Innocent, the Eager, and the Doomed.* New York, NY: Pantheon Books, 1983.

Gould, Jay R., and Wayne A. Losano. *Opportunities in Technical Writing Careers.* Chicago, IL: NTC/Contemporary Publishing Group, Inc., VGM Career Horizons, 2000.

Hackos, JoAnn T. *Managing Your Documentation Projects.* New York, NY: John Wiley & Sons, 1994.

Hackos, JoAnn T., and Dawn M. Stevens. *Standards for Online Communication: Publishing Information for the Internet/World Wide Web/Help Systems/Corporate Intranets.* New York, NY: John Wiley & Sons, Inc., 1997.

Haramundanis, Katherine. *The Art of Technical Documentation, 2nd ed.* Boston, MA: Digital Press, 1998.

Hargis, Gretchen, and others. *Developing Quality Technical Information: A Handbook for Writers and Editors.* Upper Saddle River, NJ: Prentice Hall PTR, 1998.

Horton, William. *Designing and Writing Online Documentation, 2nd ed.* New York, NY: John Wiley & Sons, 1994.

Kent, Peter. *Making Money in Technical Writing: Turn Your Writing Skills into $100,000 a Year.* New York, NY: Macmillan General Reference USA, 1998.

Matthies, Leslie A. *The New Playscript Procedure: Management Tool for Action.* Stamford, CT: Office Publications, Inc., 1977.

Microsoft Corporation. *Microsoft Manual of Style for Technical Publications, 2nd ed.* Redmond, WA: Microsoft Press, 1998.

———. *Microsoft Press Computer Dictionary, 3rd ed.* Redmond, WA: Microsoft Press, 1997.

Price, Jonathan, and Henry Korman. *How to Communicate Technical Information: A Handbook of Software and Hardware Documentation.* Berkeley, CA: Addison-Wesley, WordPlay Communications, 1993.

Rubin, Jeffrey. *Handbook of Usability Testing: How to Plan, Design, and Conduct Effective Tests.* New York, NY: John Wiley & Sons, Inc., 1994.

Strunk, William F., and E.B. White. *The Elements of Style, 3rd ed.* New York, NY: Macmillan Publishing Co., Inc., 1979.

Taormina, Tom. *Assessing ISO 9000 for Your Business: Key Elements and Strategies.* Rockville, MD: ABS Group, Inc., Government Institutes, 1999.

Tapscott, Don. *Growing Up Digital: The Rise of the Net Generation.* New York, NY: McGraw-Hill, 1998.

University of Chicago Press. *The Chicago Manual of Style, 14th ed.* Chicago, IL: University of Chicago Press, 1993.

Warriner, John E. and others. *English Grammar and Composition, 11 rev. ed. with supplement.* New York, NY: Harcourt, Brace & World, Inc., 1969.

Williams, Robin. *The Non-Designer's Design Book: Design and Typographic Principles for the Visual Novice.* Berkeley, CA: Peachpit Press, 1994.

Winter, Barbara J. *Making a Living Without a Job: Winning Ways for Creating Work That You Love.* New York, NY: Bantam Books, 1993.

Zeleznik, Julie M., Philippa J. Benson, and Rebecca E. Burnett. *Technical Writing: What It Is & How to Do It.* New York, NY: Learning Express, LLC, 1999.

Professional Organizations and Web Sites

About.com

www.telecommuting.about.com/smallbusiness/telecommuting/mbody.htm

Information about telecommuting.

http://techwriting.about.com/careers/techwriting/

Information about technical writing jobs and careers.

http://jobsearch.about.com/careers/jobsearch/cs/internationaljobs/index.htm

Information about overseas jobs.

AECMA Simplified English

www.aecma.org/Publications/SEnglish/senglish.htm

Information about the aerospace industry's controlled language, Simplified English.

Alliance for Computers and Writing (ACW)

http://english.ttu.edu/acw/

National nonprofit organization that supports teachers at all levels in using computers to teach writing.

American Chemical Society (ACS)

www.acs.org/

A self-governed individual membership organization for chemists, chemical engineers, and technicians.

American Medical Writers Association

www.amwa.org/

A group formed to promote excellence in writing, editing, and producing printed and electronic biomedical communications.

American National Standards Institute (ANSI)

www.ansi.org/

The administrator and coordinator of the United States private sector voluntary standardization system.

American Society for Information Science (ASIS)

www.asis.org/

A society focused on theories, techniques, and technologies to improve access to information.

American Society of Indexers (ASI)

www.asindexing.org/

Society for those engaged in indexing as a profession.

Andreas Ramos Homepage

www.andreas.com

Information about tech writing and job searching.

Association for Computing Machinery (ACM)

www.acm.org/

The world's first educational and scientific computing society.

Association of Teachers of Technical Writing

www.acm.org/

Professional society for teachers whose subject is technical writing.

Australian Society for Technical Communication (NSW)

www.astcnsw.org.au/

Australian society corresponding to the STC, for those engaged in technical communication as a profession.

Australian Society of Indexers

www.zeta.org.au/~aussi/

Australian society for those engaged in indexing as a profession.

Betsy's Technical Writing Site

www.frii.com/~bpfister

A good place to go for frequently asked tech writing questions, community with other tech writers, and regular articles about tech writing, as well as links to other tech writing–related sites.

British Computer Society (BCS)

www.bcs.org.uk/

The only chartered professional institution for the field of information systems engineering, the BCS exists to provide service and support to the IS community, including individual practitioners, employers of IS staff, and the general public.

Business Know-How.Com

www.businessknowhow.com

Information about home offices, small businesses, and careers.

Computer Press Association (CPA)

www.computerpress.org/

Promotes excellence in computer journalism.

High-Tech Jobs Online

www.dice.com

Very popular and extensive job site that focuses on positions in the high-tech industries or that require high-tech training and experience (for example, familiarity with global positioning systems).

HTML Writers Guild

www.hwg.org/

The world's largest international organization of Web authors with over 118,000 members in more than 150 nations worldwide, this community provides resources, support, representation, and education for Web authors at all skill levels.

IEEE Professional Communication Society (PCS)

www.ieee.org/organizations/society/pcs/pcsindex.html

Helps engineers and technical communicators develop skills in written and oral presentation.

Indexing and Abstracting Society of Canada

http://tornade.ere.umontreal.ca/~turner/iasc/home.html

Canadian society for those engaged in indexing and abstracting as a profession.

Institute of Scientific and Technical Communicators (ISTC)

www.istc.org.uk/

Largest UK body representing professionals communicators and information design-ers; its mission is to set and improve standards for communication of the scientific and technical information that support products, services, or business.

InteractionArchitect.com

www.interactionarchitect.com/

Knowledge base for designing usable and useful interactive systems.

International Association of Business Communicators (IABC)

www.iabc.com

International association for those engaged in business communications that may or may not be technical in nature (for example, marketing communications and trade or business newsletters).

International Telework Association & Council

www.telecommute.org/

Nonprofit organization dedicated to promoting the economic, social, and environ-mental benefits of telework; members share information about the design and imple-mentation of telework programs, the development of the worldwide telework sector, and research.

Internet Society (ISOC)

www.isoc.org/

A nonprofit, nongovernmental, international, professional membership organization focused on standards, education, and policy issues.

JavaScript.com—The Definitive JavaScript Resource

www.javascript.com/

JavaScript information, Web-based tutorials, and free downloadable scripts.

JobHunterBible.com

www.jobhuntersbible.com/

Web site for career-changers, run by Dick Bolles, author of the bestselling *What Color Is Your Parachute?*

Monster.com

www.monster.com

Another popular and extensive job site, not limited to high-tech jobs.

National Writers Union

www.nwu.org

The union for freelance writers working in U.S. markets.

Open Source Writer's Group

www.oswg.org/oswg/

A noncommercial, nonprofit organization whose primary goal is to improve the overall quality and quantity of free open-source and open-content documentation.

The Official TECHWR-L Web Site

www.techwr-l.com/techwhirl/index.php3

Web site supporting the technical communication community through the TECHWR-L mailing list, an unmoderated discussion forum for technical communication topics with over 4,900 subscribers worldwide.

Overseas Jobs

www.overseasjobs.com/

Resource for international employment and work abroad.

Personality: Character and Temperament

www.keirsey.com.

Web site for the Keirsey Temperament Sorter and Keirsey temperament theory.

The Plain Language Action Network

www.plainlanguage.gov/

A government-wide group working to improve communications from the federal government to the public. Its recommendations are very similar to those of AECMA Simplified English.

Society for Documentation Professionals

www.sdpro.org/

Promotes skill development and career growth for information design and delivery professionals by sharing information about technology and marketplace trends.

Society for Technical Communication

www.stc-va.org

Organization dedicated to advancing the arts and sciences of technical communication.

Synergistech Communications

www.synergistech.com

Web site for Synergistech Communications, a Bay Area recruiting service that specializes in technical communicators. Their Career Development Corner contains information you don't want to miss.

Technical Communicators' Forum (TC-Forum)

www.tc-forum.org/

Technical communicators' forum offered by technical communicators for their colleagues worldwide.

Technical Communicator's Resource Site

www.techcommunicators.com/

Extensive resource site for technical communicators.

Telecommuter's Digest

www.tdigest.com

Online telecommuting and resource guide.

Telecommuting Knowledge Center

www.telecommuting.org/

Online resources and information for telecommuting technologies.

Usability First

www.usabilityfirst.com/

Online guide to usability resources.

Usability Professionals' Association (UPA)

www.upassoc.org/

Professional association that provides a network and opportunities through which usability professionals can communicate and share information.

Usable Web

www.usableweb.com/

Collection of links about human factors, user interface issues, and usable design specific to the World Wide Web.

Useit.com: Jakob Nielsen's Web Site

www.useit.com/

Web usability by expert Jakob Nielsen.

User Interface Engineering

http://world.std.com/~uieweb/

User interface resources, including bibliography, conference information, jobs, and more.

The Web Standards Project

www.webstandards.org/

Coalition of Web developers and users whose mission is to stop the fragmentation of the Web by persuading browser makers to adopt standards that are in everyone's best interest.

World Wide Web Consortium (W3C)

www.w3.org/

Founded in October 1994 to develop common protocols that promote the Web's evolution and ensure interoperability.

Writing for the Web

www.useit.com/papers/webwriting/

Research on how users read on the Web and how authors should write for the Web.

Index

329